Engineering Disease Resistance in Plants using CRISPR-Cas

CRISPR genome-editing technology presents opportunities to engineer disease resistance traits in plants and improve crop quality. *Engineering Disease Resistance in Plants using CRISPR-Cas* introduces readers to the basics of CRISPR-Cas and discusses its potential uses in various fields. The book focuses on methods of developing disease-resistant crops using CRISPR-Cas–mediated plant disease resistance modification.

Comprehensively written, the author details all types and variants of the CRISPR toolkit. The book opens with information on the evolution of the CRISPR technology and follows a chronology of its development. Although the book concentrates on the use of CRISPR-Cas for disease resistance in plants, it also covers the technology's broader potential examining the history and development of other genome-editing tools.

Key Features:

- Investigates the regulatory, ethical and societal considerations while designing experiments.
- Discusses topics on disease development, control and plant defense mechanisms.
- Examines genome-editing tools, including Zinc Finger Nucleases (ZFNs) and Transcription Activator-Like Effector Nucleases (TALENs).
- Examines production technology to reduce bacterial, fungal and viral diseases.
- Provides information for users to discover ways to overcome the challenges associated with food security.

This book is a valuable resource for researchers, scientists, and undergraduate and graduate students who wish to gain a comprehensive understanding of genome-editing methods.

Engineering Disease Resistance in Plants using CRISPR-Cas

Zulqurnain Khan

CRC Press
Taylor & Francis Group
Boca Raton London New York

CRC Press is an imprint of the
Taylor & Francis Group, an **informa** business

First edition published 2023
by CRC Press
6000 Broken Sound Parkway NW, Suite 300, Boca Raton, FL 33487-2742

and by CRC Press
4 Park Square, Milton Park, Abingdon, Oxon, OX14 4RN

CRC Press is an imprint of Taylor & Francis Group, LLC

ISBN: 978-1-032-27113-2 (hbk)
ISBN: 978-1-032-27114-9 (pbk)
ISBN: 978-1-003-29141-1 (ebk)

DOI: 10.1201/b22901

Typeset in Times
by codeMantra

Contents

Foreword

On the invitation of Dr. Zulqurnain Khan, I am happy to write about his new book *Engineering Disease Resistance in Plants using CRISPR-Cas*, which is an informative book encompassing developments in CRISPR-Cas and its applications for disease resistance in plants. Dr. Khan worked in my laboratory in 2016. He is one of the pioneers in this field from his country. He has organized various workshops and seminars to introduce CRISPR technology in his country. Back in 2017, we organized a training workshop in the Center of Advanced Studies, University of Agriculture, Faisalabad, Pakistan, which may be the first hands-on training of the CRISPR designing and cloning. Dr. Khan's work was on virus resistance using CRISPR-Cas. I am happy to see that he has written a book in his field of expertise, in which he has been working for the last 10 years.

The CRISPR-Cas system, with its broad range of applications in basic and applied research, has been revolutionizing the life sciences. Apart from plant sciences, CRISPR system is also well established in medicine and health sciences. CRISPR is an efficient, robust and simple tool of genome editing, which has been utilized in bacteria, fungi, yeast, animals, plants and human cell lines. CRISPR has become a favored method for studying molecular mechanisms, fighting viral, bacterial and fungal diseases in plants, and eradicating vector-borne diseases such as zika, dengue and malaria. This book is covering the role of CRISPR as an effective tool against diseases in plants. The author has used a very comprehensive manner to discuss complex concepts and mechanisms. Starting from introductory chapters in CRISPR-Cas system and plant diseases and their mode of infection, the author discusses applications of the CRISPR system in general. Various methods of developing disease resistance in plants against bacterial, fungal and viral diseases have been elaborated providing ideas for future research as well. The review of the technology and the previous work is well-written, providing an overall story and historical timeline of the developments in the CRISPR system specifically for disease resistance. Moreover, discussion in transformation workflow and delivery methods will be very helpful for postgraduate students in designing their research experiments. Dr. Khan also included discussions of regulatory and ethical aspects of the technology, which is of prime importance when it comes to commercial applications of the technology and its products.

Unequivocally, this book will attract the attention of readers from various fields, including undergraduate and postgraduate students, faculty, researchers and industry. Work on innovative technologies is important to make it comprehensive and understandable for all stakeholders to bridge the industry–academia gap. Along with recognizing this valuable work of Dr. Khan for academic and scientific readers, I also appreciate the efforts of the publisher, CRC Press, for supporting such an innovative book project.

Prof. Caixia Gao
Principal Investigator,
State Key Laboratory of Plant Cell and Chromosome Engineering,
Institute of Genetics and Developmental Biology, CAS, Beijing, China

Preface

My journey into the realm of genome editing began with my thesis study in 2012. I was quite excited about planning my thesis study utilizing Zinc Finger Nucleases. Later, Prof. Sultan Habibullah Khan, my supervisor, suggested that I should take TALEs and TALENs as well. We began by utilizing TALEs in model plants to develop resistance to Begomoviruses. Following that, TALENs, CRISPR-Cas9 and dCas9 were employed, with technical assistance from Prof. Caixia Gao, IGDB, Chinese Academy of Sciences, Beijing, China. The research on viral resistance engineering utilizing CRISPR-multiplexed technology was completed and published in international peer-reviewed journals.

Cotton Biotechnology Laboratory (CBL), CABB, University of Agriculture, Faisalabad, Pakistan, was the site of the first TALEN cloning and transformation experiments. Dr. Aftab Ahmad has been exceedingly kind in providing technical assistance and imaginative ideas to the project. The CBL team held the first hands-on training and workshop on CRISPR technology in 2017, which drew the attention and interest of scientists working in biotechnology and molecular biology.

Scientists in Pakistan are now aware of genome-editing technology, and many initiatives are underway with the goal of finding answers to current issues, notably in agriculture. Our team, which includes Dr. M. Salman Mubarik, Dr. Sabin Aslam, Mr. M. Zubair Ghouri and others, has been using genome-editing techniques for targeted gene alterations in a variety of plants ranging from model plants to crops, vegetables and fruits. Tens of doctorate and master's students have finished their thesis study in genome editing. Prof. Sultan H. Khan is also directing the effort to establish Pakistan's first Genome Editing Center at the University of Agriculture in Faisalabad.

I am thankful to my team and foreign colleagues from UCDAVIS in the United States and CAS in China for their assistance and advice in building genome-editing research and facilities in Pakistan. Furthermore, I am grateful to my doctoral and master's students for lending a hand throughout this venture.

This book will cover the fundamentals of genome-editing technology, including CRISPR-Cas, and its applications for a variety of objectives. The book is aimed primarily for master's students to comprehensively increase their grasp of genome-editing techniques. Although the book focuses on the use of CRISPR-Cas for disease resistance in plants, it also discusses wider applications of the technology.

Finally, I want to express my gratitude to the publisher CRC Press and its team for their assistance with this book endeavor. Indeed, it will benefit students, academics and industry, in the advancement of CRISPR technology. The use of new solutions to tackle challenges, particularly food security and safety, is critical. I believe that the future of CRISPR-edited organisms is bright, and that a major revolution will reshape the whole landscape of agriculture and agro-based business.

Author

Dr. Zulqurnain Khan is an Assistant Professor at the Institute of Plant Breeding and Biotechnology (IPBB), MNS University of Agriculture, Multan (MNSUAM), Multan, Pakistan. Dr. Khan earned his PhD in Biotechnology in 2017 from the University of Agriculture, Faisalabad (UAF) in Pakistan. He was Pakistan's first PhD candidate with a thesis on genome editing. He has been involved in the field of genome editing since 2012. His study focuses on employing genome-editing techniques (TALEs, TALENs, Cas9, dCas9 and multiplexed CRISPR-Cas9) in model plants and cotton to develop resistance to Begomoviruses. He is also employing CRISPR-Cas9 technology to enhance cereal crop genetics for abiotic and biotic stress tolerance. He has edited two books on genome editing, one book on cotton breeding, and authored more than twenty book chapters. In addition, he has had several review and research papers published in international peer-reviewed journals. Dr. Khan was awarded with PhD scholarship and IRSIP fellowship by the Higher Education Commission (HEC) of Pakistan. He worked as a visiting researcher at Professor Caixia Gao's group at the Institute of Genetics and Developmental Biology (IGDB), Chinese Academy of Sciences (CAS), in Beijing, China. Since 2018, Dr. Khan has been teaching several courses on biotechnology, breeding, molecular biology, genetics and genome editing.

1 History and Development of CRISPR-Cas

A Bacterial Defense System

INTRODUCTION

In the genome of *Escherichia coli*, several repeated nucleotide sequences were discovered by microbiologists in 1987, linked with distinct sequences known as clustered regularly interspaced short palindromic repeats [CRISPR]. Different CRISPR types for various bacteria or archaea have been identified; however, it took approximately 20 years to identify spacers from invading DNA (Ishino *et al.* 1987). The spacer nucleotide coincided with bacteriophage DNA, indicating a memory of earlier infections (Bolotin *et al.* 2005). CRISPR sites are used to transcribe RNAs, which are then converted into CRISPR-Cas-derived RNAs, for improved and accurate identification and to break virus DNA apart [crRNA]. There are five main categories of CRISPR-Cas based on bacterial immune system. Characterizing additional unclassified CRISPR variations is now necessary (Makarova *et al.* 2015). CRISPR-Cas system is described in Figure 1.1.

The mechanism based on several CRISPR-associated proteins aids in the acquisition of new spacers, crRNA proliferation and the elimination of pathogenic nucleic acid sequences. A CRISPR-Cas system containing not only Cas endonuclease, crRNA, but also trans-activating RNA [tracrRNA] is needed to create a selective double-strand breakage [DSB] for genome engineering. To modify a desired gene, a target sequence and a cloning technique are necessary, since both RNAs have been joined to produce a single guide RNA [sgRNA] (Jinek *et al.* 2012). The CRISPR technology now being used by scientists only requires one gRNA and one Cas protein, which is sufficient to target a sequence. For targeted genome editing use of CRISPR-Cas system is effective in producing DSBs at the target region based on NHEJ and HDR—two cellular repair pathways. Target sequence selection is limited because it needs a protospacer adjacent motif (PAM), which is determined by the species from which the Cas protein is derived. When using Cas from *Streptococcus pyogenes*, 'NGG' PAM is needed at the 3′ ends of the target in conventional applications. NGG sequences are prevalent in all coding sequences hence, it is not a major issue.

As a result, now we practically can use the CRISPR-Cas system in all organisms to modify their genomes. Direct gene transfer into plant cells may also be used to produce transient expression as stable plant translation. It may be done by injecting DNA microparticles, polyethylene glycol [PEG], micropropagation and electroporation into the protoplasts (Altpeter *et al.* 2005). All of these techniques used on plants do not rely on agrobacterium-mediated transformation. As a result, no additional vectors are required for targeted editing; only the template for sgRNA and Cas are required for genome-editing. In 2013, the first findings of CRISPR-Cas-based plant genome

DOI: 10.1201/b22901-1

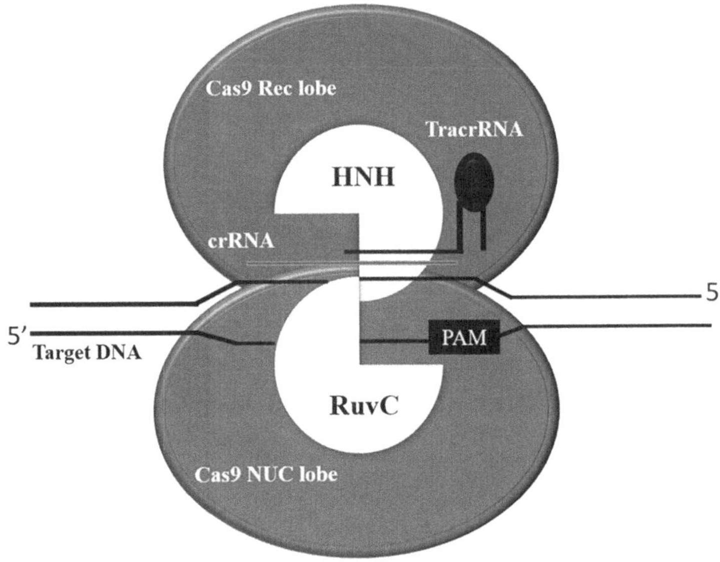

FIGURE 1.1 CrRNA, TracrRNA and Cas9 complex shown in the figure are responsible for targeting the attacking viruses in the bacteria. Cas9 proteins have two lobes each having a nuclease domain (HNH and RuvC) responsible for cutting the double strand of the DNA. TracrRNA makes a duplex with crRNA, which scans the target site for complementary bases to recruit the Cas9 nuclease for creating DSB at the target site. PAM region is a protospacer adjacent motif, which is needed to target the specific site in the DNA.

editing were released (Nekrasov *et al.* 2013; Shan *et al.* 2013; Feng *et al.* 2013; Li *et al.* 2013; Xie & Yang, 2013). Using the nonhomologous end-joining [NHEJ] or the homology direct repair [HDR] method, the system may be used to repair damaged DNA. When infected with viruses, bacteria incorporate smaller segments of the viral DNA into their own genome in a specific pattern typically known as CRISPR arrays, which is implicated in the recognition and elimination of viruses following subsequent infection. The general mechanism behind the capability of CRISPR Cas9 to target viral DNA involves the generation of RNA segments from CRISPR arrays which ultimately identify and bind with specific regions of viruses' DNA. The bacteria then utilizes Cas9 protein or similar enzyme to introduce a cut into the DNA with the consequential disabling of viruses. These processes have recently been improved to function as a more sophisticated tool for molecular biology and genetic engineering.

HISTORY OF CRISPR-CAS

In 1987, microbiologists found several repetitive sequences in the *E. coli* genome that were linked with distinctive sequences known as clustered regularly interspaced short palindromic repeats (CRISPR). They discovered a novel genetic makeup consisting of alternating repeat and non-repeat nucleotide sequences similar to the sequence of that gene, although its molecular significance was not yet known. The goal of these amazing mechanisms could not be fully understood for 20 years (Ishino *et al.* 2018). Experimental data proved CRISPR to be a key component of the bacterial defense mechanism against bacteriophage infection. Two distinct CRISPR loci were found in separate strains of *Streptococcus thermophilus* by researchers (Morange *et al.* 2015). The CRISPR system spacer sequences were sequenced, and it was discovered that they matched a number of bacteriophage or plasmid sequences exactly (Xue *et al.* 2015). This gave rise to the hypothesis that CRISPR-Cas was a bacterial defense system against encroaching species (Mojica & Rodriguez-Valera, 2016).

A wild type [WT] strain of *S. thermophilus* was exposed to two separate, highly virulent bacteriophages in order to test this theory. Nine distinct phage-resistant *S. thermophilus* strains were produced consequently, and further examination of the CRISPR loci in these mutant strains revealed that additional spacers had been introduced adjacent to those of the WT strain. Furthermore, the new spacers' sequences matched those found in the phages' genomes that were employed during the test (Charpentier & Marraffini, 2014). This supported the theory that bacteria under virus stress may incorporate fresh spacers from the bacteriophage genome sequence through CRISPR-Cas, which may result in a variety of phenotypes of bacteria resistant to phages. During the investigation, researchers noticed that a certain subset of CRISPR-Cas-associated genes that coded for Cas proteins were located near CRISPR sites. These genes were similarly important for immunity using CRISPR, because their suppression impairs CRISPR function (Peng *et al.* 2016; Marraffini & Sontheimer, 2010). Timeline of CRISPR-Cas is described in Figure 1.2.

CRISPR-CAS: ADAPTIVE IMMUNITY OF PROKARYOTES

Prokaryotic organisms are intimidated by a large number of viruses and have developed numerous defense strategies. CRISPR-Cas system provides adaptive immunity of bacteria from the invading viruses. A spacer, a brief segment of the viral genome, is incorporated into the CRISPR-Cas locus during viral infection to immunize the host cell. Spacers are translated into tiny RNA which guide Cas nucleases to cleave the viral genome. The complete mechanism of natural CRISPR defense in bacteria is described in Figure 1.3.

STEPS OF ADAPTIVE IMMUNITY IN PROKARYOTES

ADAPTATION

Adaptation or acquisition is the initial stage in CRISPR-Cas-mediated defense. In this stage, the CRISPR-Cas system incorporates the phage genetic material, so the organism has a way to detect and change its defenses against invasion through that phage variant in the future. All CRISPR-Cas systems include the necessary Cas 1 or Cas 2 proteins, regardless of the type, and they facilitate the adaptation process. Since Cas 1 or Cas 2 expression alone does not enhance spacer acquisition, both proteins are necessary for this stage. Further investigation showed that genetic variations of either of the *cas*1 and *cas*2 genes in *E. coli* CRISPR segments prevented spacer acquisition, assisting the important roles of both proteins in spacer acquisition. It was also discovered that upregulation of both *cas*1, as well as *cas2*, enhanced spacer incorporation (Marraffini *et al.* 2015).

In *E. coli*, one of the most researched adaptation systems, two Cas1 dimers and one Cas2 dimers combine to form a symmetrical heterohexameric protein complex. Although Cas2 nuclease activity is not necessary for spacer acquisition, neither of the proteins lacks

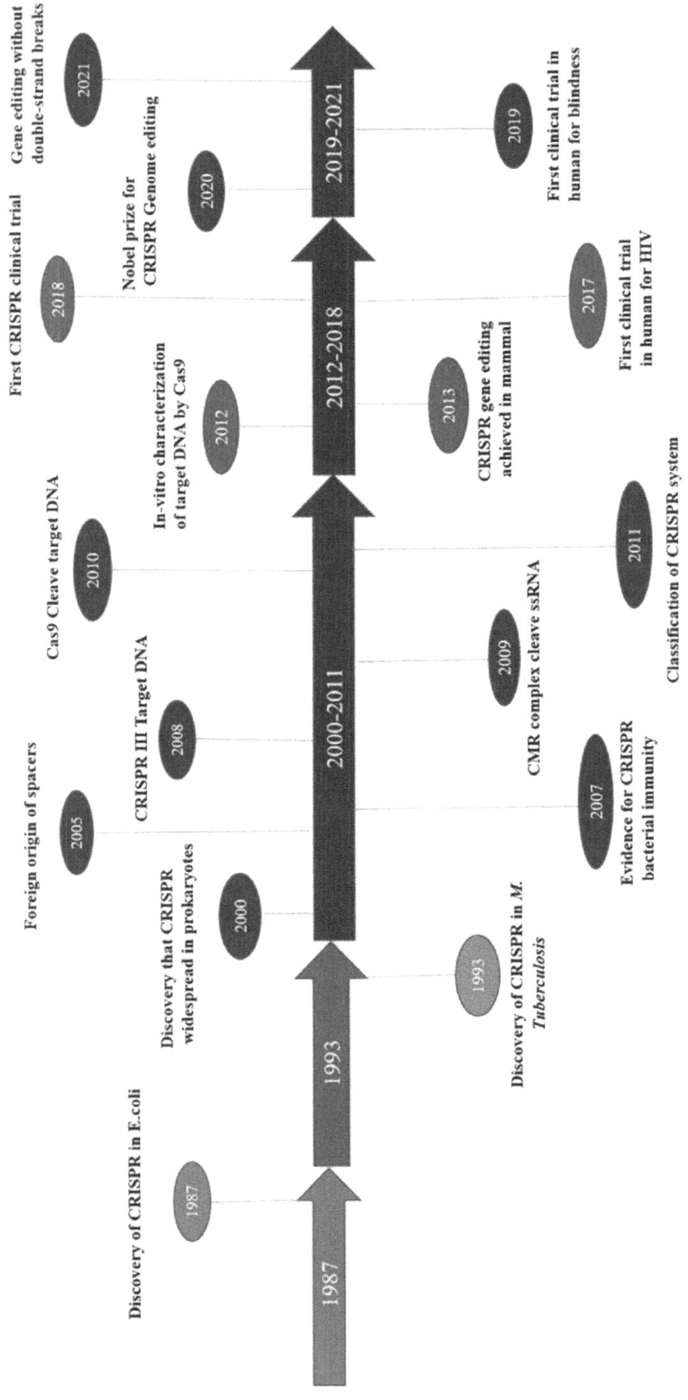

FIGURE 1.2 Timeline of CRISPR-Cas.

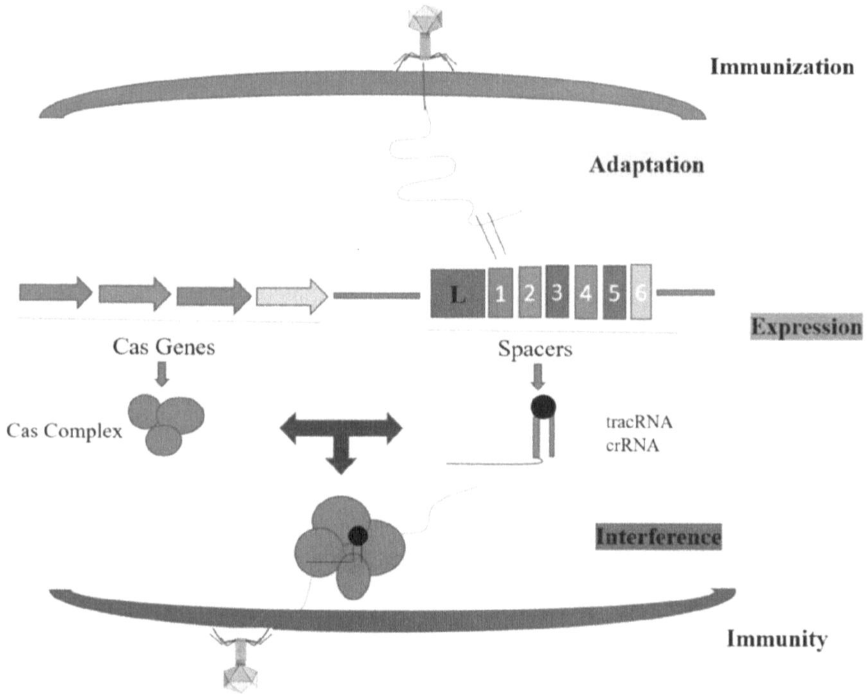

FIGURE 1.3 Natural CRISPR-Cas prokaryotic adaptive immune system. The key steps of CRISPR-Cas immunity. (1) Adaptation: insertion of new spacers into the CRISPR locus. (2) Expression: transcription of the CRISPR locus and processing of CRISPR RNA. (3) Interference: detection and degradation of mobile genetic elements by CRISPR RNA and Cas protein.

nuclease activity. The Cas1–Cas2 complex performs two functions during the adaptation step: it removes protospacer DNA [a stretch of foreign DNA that comes before the spacer sequence] and incorporates it into CRISPR-Cas sequence. PAMs which are small [2–5 nucleotides] sequences that are specific to each bacteria and CRISPR-Cas subtype is involved in the determination of spacer alignment either in class 1 or class II systems within CRISPR array (Van der Oost *et al.* 2009). The ssDNA is where Cas1 attaches in the type I system form of the PAM complementary sequence. Cas can detect the NGG sequence in its double-stranded form in a type II system. Additionally, PAM seems to participate in both the self-/nonself-differentiation and the interfering phase.

The Cas1 and Cas2 subunits of the CRISPR system found in *E. coli* type I-E first recognize PAM complementary sequences in ssDNA to initiate spacer acquisition. The core Tyrosine residue [Tyr22] in Cas 1 subunit limits the size of the double-strand DNA domain of protospacer at 23 nucleotides, which functions as a ruler to frame the foreign genetic material. The core DNA region is stabilized by the Cas 2 dimer of complex Cas1–Cas2. From 3′ end of the core duplex area, two ssDNA strands with an overhang of at least seven are present. The PAM complementary sequence corresponds to the last three nucleotides (Van der Oost *et al.* 2009). These three nucleotides are cut by the nCas1 regions, yielding two 3′-OH compounds and a developed protospacer.

The Casl and Cas2 complex integration activity facilitates the protospacer DNA incorporation into CRISPR-Cas array. The protospacer intermediate 3′-OH groups catalyze two successive nucleophilic assaults on the first repetition of the CRISPR array at both of its 5′ ends. The end outcome is a larger CRISPR array with enzymes that subsequently repair two imperfect single-stranded DNA repeats, adding a new spacer among them. The most recent spacer to be acquired is the first one on the CRISPR array because of the bias in favor of the repetition closer to the leader sequences during selection.

The adaptation stage in certain type I, type II and type V systems also depends on the action of the Cas 4 nuclease. Since Casl is connected to reverse transcriptase, spacers from RNA-based invaders may be acquired and then reverse-transcribed into DNA using *Marinomonas mediterranea*'s type III-B Cas system (Patterson *et al.* 2017). The feature of such a complex is assumed to be equivalent to the well-known adaptation mechanisms of type I and type II systems in CRISPR-Cas systems, despite Casl and Cas2 being widespread, including almost in CRISPR systems.

Expression and Processing of CRISPR

Subtype-specific procedures and enzymes are used in the transcription, and small crRNAs are created by processing the CRISPR-Cas array and *cas* genes. All CRISPR-Cas systems convert the CRISPR region into crRNA precursors [pre-crRNA] (Barrangou & Marraffini, 2014; Carte *et al.* 2014). Cas proteins or cellular ribonucleases then cleave and process the pre-crRNA to produce maturing crRNA smaller subunits. The only spacer sequence found in this matured crRNA is present and repeats region fragments on each side (Barrangou & Marraffini, 2014). The enzymes that transform pre-crRNA into complete crRNA fragment in type I systems is called Cas6 [formerly known as Cse3]. For instance, type I-E spacers following transcription take on a stem–loop structure that is recognized and cut by Cas6e. After cleavage, this type of I-E-specific protein is still anchored to the 3′ end of the crRNA (Chevallereau *et al.* 2019).

Type IV CRISPR-Cas systems are uncommon or lack the common proteins Casl, Cas2 and Cas4 that are required for adaptation and cleavage (Terns & Terns, 2011). Contrary to other class one systems, type IV lacks Cas6—a protein required for the conversion of pre-crRNA into mature crRNA in types I and III. The three proteins that make up its multisubunit effector module are Cas5, Cas7 and Csf1—the system's distinguishing protein (Koonin & Makarova, 2009). To fully understand its methods of adaptability and bacterial immunity, type IV provides more insights (Wiedenheft *et al.* 2012).

Type II systems depend on the small trans-activating RNA molecule, RNase III and Cas proteins, instead of carrying the *Cas*6 gene [tracrRNA]. Three stem–loop hairpin structures are present in tracrRNA, which complements the repetition pattern (Barrangou & Marraffini, 2014). Following transcription, tracrRNA attaches to pre-crRNA molecules that are made of alternating sgRNA repeats and sgRNA spacers. So, as a molecular anchorage, Cas maintains the tracrRNA–pre-crRNA connection for RNase III to recognize and cleave pre-crRNA later on for full processing.

The proteins Casl2 and Casl3, respectively, convert in type V and type VI; pre-crRNA develops into a matured crRNA, except the aid of tracrRNA molecules (Schindele *et al.* 2018). Pre-crRNA monomers might be used as target breakage guides; therefore, the processing step is not necessary for subtype VI-A. For full crRNA processing, both type II and type III systems need an additional trimming phase using a methodology based on a ruler (Belhaj *et al.* 2015). For system of type II, trimming happens at the 5′prime end, but it happens at the 3′ prime end in the type III systems.

Interference

For the purpose of interference, active ribonucleoprotein surveillance complexes formed as a result of interaction between cas proteins and crRNA can successfully repel invading foreign nucleic acids via sequence specific cleavage, thereby preventing proliferation and propagation of unwanted foreign DNA (Van der Oost *et al.* 2009). Type 1 systems produce a CRISPR-Cas complex called the Cascade pathway for antiviral resistance. It has multi-protein backbone that connects the crRNA molecules to a number of Cas proteins subunits. The Cascade complex enable the mating of crRNA also with homolog invasive DNA strand, which identifies that the DNA is unwound, and the invader's molecule binds to the PAM site there. This pairing causes the development of an R-loop with three strands, which in turn causes Cas3—the distinctive protein of type I systems—to be recruited (Černý & Střížˇ, 2019; Barrangou & Marraffini, 2014). The single-stranded DNA strand that is not connected to the Cascade complexes is cut by Cas3 (Deretic *et al.* 2005).

Additional cellular endonuclease or the previously described Cas3 nuclease activity not reliant on a cascade might cause complete destruction. Although Cas6 is missing from both the complexes, type III system are comparable to type I in that they both rely on Cascade complexes comprised of several proteins and crRNA. Types III-A and III-B are distinct from other interference mechanisms, because they target both RNA and DNA through Cas10 protein. Since the Cascade complex attaches to a developing single-stranded RNA transcript, type III systems interfere while the target DNA is transcribing. In intervals of 6 nucleotides, this interaction allows Cas10 to cleave the corresponding DNA duplex and Cas7 to direct the cleavage of single-stranded RNA molecule (Westra *et al.* 2014). Recent research suggests that Cas10 protein has a further function in the activation of nonspecific using RNase Csm6, converting ATP molecules into cyclic oligoadenylates. These oligoadenylates activate Csm6, which, while not being a component of the Cascade effector complex, performs an additional function by non-specifically foreigner transcripts.

Type II CRISPR-Cas systems depend on Cas or tracrRNA for interference, similar to the biogenesis stage (Sampson & Weiss, 2014; Shmakov *et al.* 2015). In interference, the RNAs crRNA and tracrRNA function as an endonuclease led by Cas. They couple up because tracrRNA complements using the spacer segments carried with crRNA, a dual RNA complexes are produced. Which then binds to Cas protein and causes conformational changes that cause Cas to become activated. When activated, the combination of guide RNA explore for the proper PAM site, opposite to the target strand, by screening foreign genetic elements (Collias & Beisel, 2021). Once the DNA has been located, it is uncoiled, and the target ssDNA along with crRNA, leading in an R-loop form induces DSB at the three nucleotides upstream of PAM region as a result of coordinated maneuver between HNH and RuvC catalytic domains (Barrangou *et al.* 2013).

For subtypes, 5-A, V-B and V-C, respectively, Cas12a [formerly Cpf1], Cas12b and Cas12c are required by type V CRISPR systems (Hille *et al.* 2018). According to phylogenetic study and their shared bilobed shape, these proteins are somewhat related to Cas. Cas12a or Cas12b asymmetrical cleavage of the DNA duplexes on both sides, following crRNA attachment to the target sequence and PAM region recognition, results in staggered breakage with 5- or 7-nt overhangs on Cas12a or Cas12b, correspondingly. But the interference method of Cas12a differs from that of Cas and Cas12b in that it only depends on crRNA for effective cleavage, not tracrRNA. Cas12c still needs to investigate its structure and actions further. The Cas13 protein presence characterizes that the newly identified type VI All RNase have binding sites present in this protein.

The RNA cleaving capacity of Cas13 with its crRNA sequences can be utilized for RNA detection, imaging and regulation and this ability is supported by nonspecific cleavage of other sgRNA, much like the Csm6 enzyme in type III systems. Preferably, cleavage takes place before uridine [U] residues (Cox *et al.* 2017). Species-dependent protospacer flanking site is additionally crucial for Cas13 protein activation, similar to PAM in DNA target. Cas13 undergoes conformational changes upon binding to crRNA, which facilitates ssRNA pairing. Cas-13 activation driven by HEPN's (higher-eukaryotes-and-prokaryotes nucleotide-binding) and the resulting composite Rnase active site is capable to generate both cis and collateral cleavage of RNA molecule.

CLASSIFICATION OF CRISPR-CAS

The categorization of the CRISPR-Cas system into two primary classes, which are further subdivided into six subtypes, is based on the variance interference complexes. Class 1 includes types I, III and IV; all other types fall under a different class (Mohanraju *et al.* 2016) (Figure 1.4).

Class 1 system used many components, while class 2 system used ribonucleoprotein [RNP] complex, crRNA and just a single protein to fight intruders (Shmakov *et al.* 2015). The majority of the CRISPR-Cas systems that have been characterized so far identify a short sequence of nucleotides termed a protospacer or NGG at the 5′ prime end that interact with genetic elements (Wang *et al.* 2016; Yamano *et al.* 2017). The type I system multisubunit RNP complex, also known as a CRISPR-Cas complex, is employed to identify and cut DNA targets (Brouns *et al.* 2008).

Most Cas5d for subtype I-C in type I system used Cas6 to detect or work with RNA with CRISPR repeats (Carte *et al.* 2008). The Cascade is guided by the PAM [NGG]

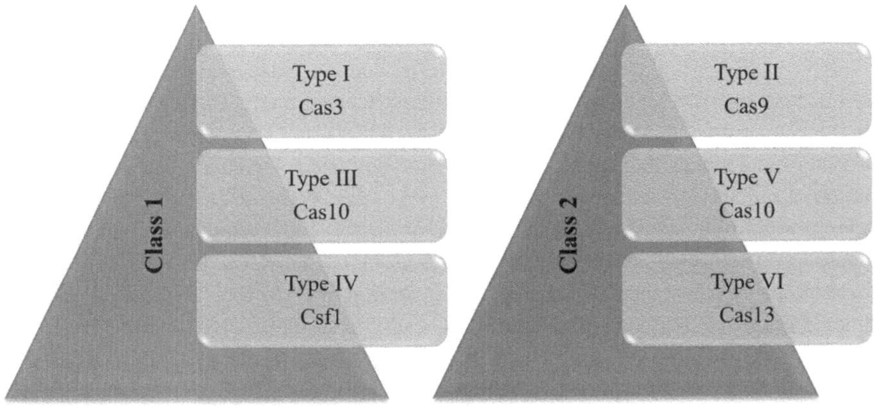

FIGURE 1.4 CRISPR-Cas classes and types.

sequence, which activates the Cas3 restriction enzymes and helps it detect the sequence of DNA (Hayes *et al.* 2016). However, the type III system, which mimics the type I combinations, use crRNA that may be coupled to a multiprotein complex to build a viral defense that targets either DNA or RNA substrates. Following the backbone targeted transcript of the Cas10 subunit, seed motifs are employed to attack the rear-end small molecule Cmrl to target the RNA that was present at the 3′end of crRNA (Pan *et al.* 2019; Li *et al.* 2018). The nuclear breakage domain of Cas10 subunits HD [Histidine] performs DNA/RNA cleavage. In order to prevent viral diseases, cyclic oligonucleotides [COA] not only systematize the RNP complex but also CARF [CRISPR-associated Rossmann fold] ribonucleases [Csm6/Csx1 families] domain transcription factors. COA is based on the synthase activity that uses the Cas10 domain and is catalyzed by a similar targeted RNA (Wang *et al.* 2013; You *et al.* 2019).

Class 1 uses type IV interference restriction enzymes such as Cas3 and Cas10 as adaptation modules (Koonin *et al.* 2017). While there is some information on crRNA maturation and RNP complex formation, it is still difficult to understand type IV's activities. However, in Class II, types II, V and VI are employed together with the protein to create a complex of RNP that serves as a defense against intruders. While tracrRNA and RNase III are actively employed for the creation of crRNA, not only type Cas12a but also Cas13 may process the crRNA itself for CRISPR RNA maturation (Fonfara *et al.* 2014). Contrary to type II Cas, which uses a variety of non-targeted stranded PAMs to induce blunt DSBs, Cas12a and Cas12b activity rely on the recognition of T-rich PAM sequences, after which DSB cleavage takes place (Yang *et al.* 2016). The type VI system, which cleaves ssRNA, degrades them and then targets the RNA that is complementary to the crRNA, encoding the Cas13 protein. The Cas13a protein has been used to identify pathogens. In addition, when bound as an activator via a complementary target of ssDNA, RNP complexes of Cas12a and Cas12b may accelerated the cleavage of nonspecific ssDNA.

VARIANTS OF CRISPR-CAS

Researchers have worked to enhance using CRISPR systems to improve their specificity, efficacy and consistency, despite the aforementioned drawbacks.

Cas9

The NHEJ pathway only serves to prevent the required HDR process, repairing the DSB when gene knockouts are not necessary. As previously mentioned, a Cas nickase variant [Casn] is produced by adding a particular mutation into the RuvC domain of Cas. Instead of creating double-stranded breaks [DSB], Casn nicks the target DNA. Since base excision repair is the preferred method for fixing single DNA nicks, you may utilize Casn to increase the effectiveness of the procedure by lowering the incidence of indel mutations brought on by unintended NHEJ repairs. The efficiency of Cas-directed genetic modification has to be further improved; nickases can be used. A group of Casn targeting opposing strands when nearby guide RNA targets are displaced by a specific number of nucleotides is known as a double nicking method. The combination of Casn systems produces double-strand break with overhangs described by gRNA, which, when combined with HDR, can result in precise gene alterations, or when paired with NHEJ, can start precise deletions in crucial alleles. Double nicking significantly boosts specificity because, unlike wild-type Cas, it can minimize, off-target effects as double-strand breaks can lead towards undesirable genetic variations when repaired by the nonhomologous end-joining pathway, when compared to the results which is swiftly corrected through greater base excision repair even if one of the Casn continues to act off-target (Koonin & Makarova, 2019). A disadvantage of this approach is that it necessitates simultaneous creation and delivery of two different guide RNA molecules.

Cas12a [*Cpf1*]

As previously established, type V Cas12a is categorized as a Class 2 CRISPR system, since it is comparable to Cas in that it simply relies on RNA molecule to generate double-strand break. Instead of the dual crRNA: tracrRNA guide used by Cas, it only needs a crRNA molecule to direct it to its target, and the resultant DSBs are staggered cuts with 5-nt 5′-overhangs rather than the blunt cuts produced by Cas. Additionally, while Cas enzymes recognize PAMs with a high G content, Cas12a prefers to link to targets with a high T content, and this range of recognized PAMs has recently expanded as a result of engineered Cas12a variants. Multiplex genome editing is made possible by the delivery of a single pre-crRNA framework to the cell, that is subsequently broken down by Cas12a in to other different crRNA particles that spectrally selective genes (Makarova *et al.* 2011). In addition, Cas12a can process all its crRNA though the RNase III activity, which reduces off-targeting effects and gives it a benefit over Cas. The accessory cleavage activity of Cas12a can be helpful as a novel tool in modifying and targeting DNA via HDR pathway, because staggered breaks are selectively repaired by this process instead of NHEJ AsCpf1 and LbCpf1—two Cas12a variants from the Lachnospiraceae bacterium ND2006 and *Acidaminococcus* sp. BV3l6, respectively, exhibit comparable on-target efficacy to SpCas in human cells (Bondy-Denomy *et al.* 2015).

Cas13a [C2c2]

In comparison to its siblings, the newest member of the CRISPR family is particularly distinctive. Despite being a class 2 CRISPR system, type VI Cas13a can only cleave RNA thanks to the two HEPN sites that are active, unlike Cas and Cas12a can also cleave DNA. It has a similar ability to Cas12a to process its own crRNA, allowing it to target multiple positions with a single pre-crRNA template. Similar to RNA silencing techniques based on RNA interference, Cas13a's RNA-cleaving abilities may not only be utilized as post-transcriptional suppression but also cleaves nuclear transcripts and has higher specificity than RNA interference (RNAi). Single DNA sequences are translated into a variety of splicing variants during transcription. Due to alternative splicing, when DNA is targeted using Cas13a According to CRISPR-Cas systems, most mRNA isoforms were impacted. By employing Cas13a, it is feasible to target just a particular isoform to examine without interfering with the action of other variants, its function and effect. Additionally, Cas13a may target pre-mRNA, which is useful in repairing the defects brought on by faulty splicing, since the enzymes may act before the wrong splicing starts (Van der Oost *et al.* 2014). The random RNA-cleaving ability of Cas13a may limit its effectiveness to just a therapeutic approach.

dCas

The modification of Cas systems that are being used as tools to control gene expression has previously been addressed in greater depth. The deactivated Cas system (dCas) loses its ability to cleave DNA but maintains the capacity to attach to specific regions when multiple silencing alterations change the RuvC or HNH catalytic domains. According to research, this dCas version of the Cas enzyme may block transcription on its own (Tadic *et al.* 2019). This is probably achieved by either interrupting the extension mechanism, if the target molecule is a component of an open recognition region, or by blocking the RNA polymerase from partnering with the promoter sequence that dCas targets. The dCas process may be further altered in a variety of ways, such as by fusing it directly or indirectly with transcription activators [like VP64] and transcription repressors [like KRAB] to enhance the efficacy of dCas-mediated transcription inhibitory activity. A particular DNA sequence expression will be increased as a result. Because dCas does not permanently alter the genomic DNA, it only modifies the expression of genes temporarily (Koonin & Makarova, 2019). However, the combination of epigenetic modifiers with dCas allows for precise and persistent changes to genetic expression. The most current innovations and techniques for dCas-based epigenetic alteration and transcriptional regulation were gathered by Brocken et al.

eSpCas, SpCas-HF1 and HypaCas

Modifying the Cas machinery interactions with the linked DNA strand is one method for increasing CRISPR targeting specificity. Cas protein cleavage specificity could be affected by sgRNA nature, secondary structure and nucleotide preferences and alterations. Therefore, any mismatch between the above mentioned factors could impact targeting specificity with unpredictable off-target effects. After *S. pneumonia* Cas [SpCas] attaches to the target location, a stable side separation is sustained by two different kinds of interactions: the binding of gRNA to the target strand and the creation of a positively charged groove as a result of an unintended interaction between the HNH or RuvC domains (Chen *et al.* 2017). Re-hybridization between the targeting or nontargeting strand is facilitated by weakening interactions that reduce positive charge on the strand that is not the target. Two 'enhanced specificity' SpCas variants [eSpCas and eSpCas] were produced by engineering SpCas mutations that reduce the strength of groove contacts by changing a single positive-charged amino acid residue. These variations greatly reduced the amount of off-target cleavage while maintaining on-target efficacy comparable to WT SpCas.

With the creation of the high-fidelity SpCas-HF1 variant, which resulted in precise on target genome editing with greater efficacy as well as limited off-target indel generation. However, instead of destroying the connections between the nontarget strand, Kleinstiver and his colleagues changed four SpCas residues that generated hydrogen bonds with targeted strand phosphate backbone. This impaired when there are any mismatches present, gRNA binds to its DNA targets. SpCas-HF1, which was created by alanine substitutions in all four residues, along with eSpCas, also demonstrated equivalent on-target action to WT SpCas without negatively significant off-target impacts.

Single-molecule fluorescence resonance energy transfer [smFRET] is a most recent technique used by the researchers to investigate the differences between targets for SpCas-HF1 and eSpCas. Researchers have discovered that SpCas-HF1 and eSpCas bind to mismatched sequences and then halt in an inactive conformation. They also defined the roles of Cas's noncatalytic REC3 domain, which controls complementary targets or HNH catalytic properties. They created a hyperaccurate Cas variant [HypaCas] using this newly learned information by inducing mutations in the REC3 domain. When compared to SpCas-HF1, this mutant has equivalent or better selectivity with the same on-target effectiveness as WT Cas. Main catalytic sites cause Cas13 RNase activity to start when it links to the target.

CRISPR TOOLKIT

Numerous efforts have been made earlier to change the genome in plants, but CRIPSR Cas9 has proven to be the most revolutionary technology for precise genome editing as well as a next generation molecular diagnostic tool (Nekrasov *et al.* 2013; Shan *et al.* 2013). With advancement in technology, a number of very effective CRISPR-Cas vector delivery techniques, including virus infection supersession, cis-element gene destruction, genetic deletion, gene knockout and genetic modification multiplexes, have been created for plant genome editing. The Cas system can now be delivered to plant cells more precisely, accurately and specifically thanks to ongoing advancements in this editing toolkit (Figure 1.5), which have also led to the identification of additional Cas variations. Numerous species have quickly developed the CRISPR-Cas immune bacterial system into an efficient genome-editing tool. The CRISPR-Cas nuclease is led to its targeting site by a gRNA in this system. The scientists have extensively used gRNA Cas technology in plants, animals and cell lines (Bortesi & Fischer, 2015; Ran *et al.* 2013; Belhaj *et al.* 2015; Fichtner *et al.* 2014). In plants, gRNA genome engineering through CRISPR-Cas has been useful for single- and multiple-gene knockouts to targeted donor nucleotide insertions and to targeted transcriptional regulation by fusing transcriptional activation or repressor domains into an inactivated Cas (Fauser *et al.* 2014; Piatek *et al.* 2015). The use of multiplex CRISPR cas system to target multiple genomic sequences simultaneously is a noteworthy trait, as compared to the related technologies like ZFNs or TAL effectors, where the recoding of a new protein is compulsory for each target gene sequence (Beerli & Barbas, 2002).

The gRNA Cas, where a 20 nt alteration in the gRNA is enough, is opening the door for simultaneous editing and the creating intricate regulatory circuits among other engineering possibilities (Bogdanove *et al.* 2010). In rice and tobacco, direct transfection of CRISPR-Cas and gRNAs into plant protoplasts, followed through single-cell plant regeneration, has been shown to be effective for genome editing; however, the effectiveness remained relatively low, and mother plant regeneration from protoplasts is currently not possible for several crop species (Nielsen & Voigt, 2014). A possible option for plants is to use T-DNA transformation induced by an agrobacterium together with callus induction as well as organogenic plants regeneration. In

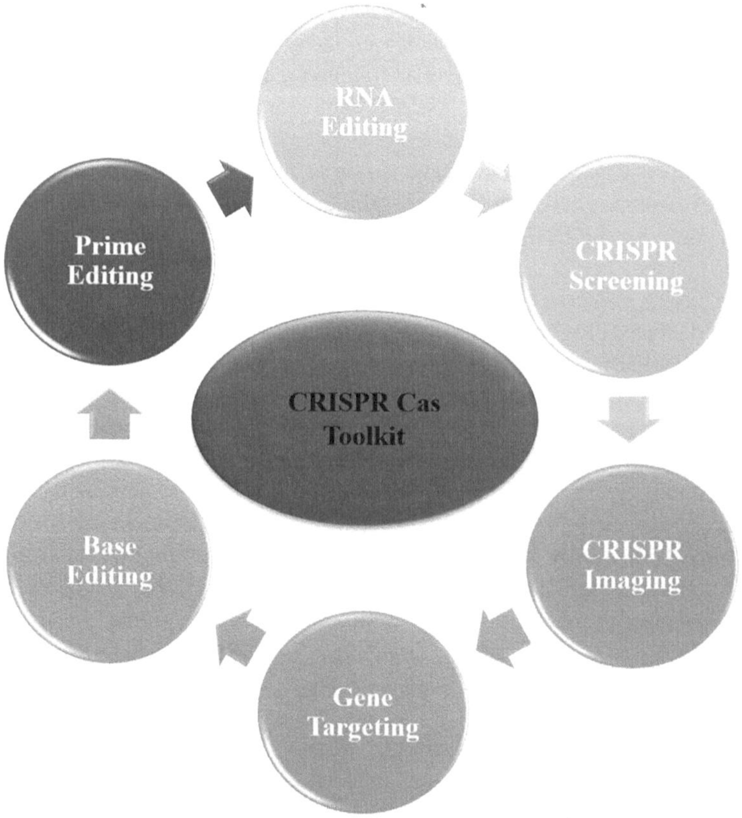

FIGURE 1.5 CRISPR-Cas toolkit.

this situation, in addition to working momentarily during callus induction, T-DNA-delivered guide RNA may also incorporate into the genome and continue to work in somatic tissues (Eeckhaut *et al.* 2013).

To fully use the T-DNA technique, it is essential to expand the capacity to combine several gRNAs with Cas inside a single T-DNA, since it has been shown that all-in-one plasmid approaches dramatically boost editing efficiency (Barrangou *et al.* 2007).

The CRISPR toolkit has been utilized frequently in the last decade, because it uses a less complex RNA-guided DNA recognition mechanism for genetic modifications. This toolkit also delivers novel methodologies, including live cell imaging, high-performance functional genomics screens and point-of-care diagnostics, which are utilized for creating desirable genetic features as well as treating genetic disorders. By instructing nuclease to attach and cut various sequences of nucleic acids, CRISPR enables gRNA adaptive protection against invading genetic elements (Marraffini & Sontheimer, 2010). Through a process termed adaptation, bacteria seize fragments of foreign genetic material and add them to their CRISPR genomic array.

By matching targeting nucleic acid base pairs, CRISPR-Cas array transcription produces crRNAs that bind to Cas nucleases and give specificity (Brouns *et al.* 2008). CRISPR-Cas has emerged as a novel technique for genome modification of any area of the genome utilizing sgRNA (Jasin & Rothstein, 2013). The targeting DNA is precisely unwound and cut using this system's two components: a chimeric sgRNA and the CRISPR-Cas with the cleavage region being chosen by complementarity with the sgRNA (Cong *et al.* 2013). In this technique, the PAM region's existence is the sole constraint on the DNA sequences that may be targeted. The CRISPR technology has shown to be very effective for site-specific genome modification. In recent years, bacterial and human cells have created the deactivating Cas protein, also known as dCas for programmed RNA-dependent DNA-binding protein (Wang *et al.* 2013). Targeting RNA polymerase binding and elongation by the nuclease-inactive Cas protein [dCas] to a gene's coding region may significantly decrease transcription. According to Jinek *et al.* 2012, dCas may be controlled to specifically recruit various protein effector DNAs for either gene activation or suppression interference. Recently, an efficient transcriptional interference was achieved by fusing dCas with the Kruppel-associated box [KRAB] repressor domain (Parsi *et al.* 2017; Gilbert *et al.* 2014). Additionally, multiplexed endogenous gene control and stable gene suppression were accomplished using CRISPRi (Kearns *et al.* 2014; Gilbert *et al.* 2013). Nontargeted gene transcription was very slightly impacted.

CRISPR-CAS SYSTEMS AND GENOME EDITING

The CRISPR technology has garnered significant attention and popularity since its development in 2012, with researchers exploring its potential in a variety of areas. Jinek and his colleagues proposed that the 3′ positions of the crRNA are connected to the 5′ positions of tracrRNA. It is possible to combine the Cas dual gRNA complexes tracrRNA:crRNA into a singular chimeric RNA in type II systems. Such a method would enable the fabrication of the hybrid RNA particle, subsequently known as sgRNA or gRNA, to cause planned DNA cleavage [gRNA] (Tyagi *et al.* 2020). A plasmid encoding the GFP gene was precisely and effectively cut by Cas including all five, as planned, of the distinct guide RNA molecules that were created to focus on the GFP gene, proving the validity of this concept (Hryhorowicz *et al.* 2017).

Soon later, other findings would reveal CRISPR's full potential as a genomic editing tool. Using the CRISPR-Cas, started in 2013, technologies are being developed for particular modifications [insertions, deletions and single-nucleotide replacements] into the genes of strains of *S. pneumoniae* or *E. coli* (Nodvig *et al.* 2015).

Type II Cas systems, as previously noted [see Interference], only use a single chimeric guide RNA molecule that can be readily created from a dual RNA complex of crRNA and tracrRNA. Guide RNA molecules include both a scaffolding segment that links to Cas protein and also targeting motif that directs the mechanism to the targeted site. When a PAM domain has been found, the target region for Cas-guide RNA, also known as seed region, will begin pairing only with target gene in the 3′–5′ direction (Tyagi *et al.* 2020). Mismatches in the seed sequence cause pairing to break down and degrade Cas cleaving activity, although it does not necessarily impair Cas performance at the 5′ PAM-distal end. A double-strand break [DSB] in the DNA is caused by homology between the gRNA and the target sequence, and is catalyzed by both of Cas catalytic domains, HNH and RuvC. The two strategies include homology-directed repairs and NHEJ through which DSBs are repaired [HDR] (Chen *et al.* 2017; Vejnar *et al.* 2016). If the nucleotides at both sides are sufficiently complementary, NHEJ is a robust and error-prone procedure wherein random DNA fragments align with both ends of the double-strand break and are connected by endogenous repair mechanisms. The primary method by which DSBs caused by Cas are fixed by NHEJ pathway, without the requirement of repair template after subsequent cleavage. Insertions and deletions of small nucleotides at DSB area may result from NHEJ, and these changes may in turn cause a wide range of frame shift mutations, reductions or addition. These mutations brought on by Cas-induced DNA double-strand breaks might be useful when trying to create a knockout inside the target gene because indels (**in**sertions and **del**etions) frequently lead to early stop codon, which turns the gene inactive (Friedland *et al.* 2013). NHEJ is not ideal for the alteration of the single base insertion in particular sequence, since it is proven to be highly unpredictable and random process.

After Cas cleavage, HDR emerges as a more accurate technique for double-strand break repair and inclusion of certain sequence (Arora & Narula, 2017). In contrast to NHEJ, HDR calls for Cas and the gRNA to be given to the cell together with the sequences in a template strand. The template ends must match the DSB's ending area on both sides for HDR to be effective. HDR efficiency could be promoted by prior delivery, use of highly specific homologous template, or either by using modified Cas 9. Due to the highest Cas activity and the much higher effectiveness of nonhomologous end joining as compared to HDR, wild-type segments, and a tiny population of the planned HDR-repaired pattern coexist in this mechanism (Gao *et al.* 2017). To increase the in vitro effectiveness of HDR, it is crucial to isolate and amplify the target sequence.

CRISPR systems have been used to alter gene expression, in addition to mutations and gene editing. Scientists created a Cas variant known as 'dead Cas' [dCas] that could still connect to DNA but lacked DNA cleaving ability by inserting two mutations into the RuvC and HNH catalytic domains. Catalytically deactivated Cas9 (dCas9) can be deployed in multiplex targeting of diverse gene promoters specified by SgRNA via fusion with transcription repressors. The targeted gene's binding hindered transcription, perhaps by sterically reducing the activity and binding of the RNA polymerase [RNAP]. Moreover, depending on the gene region that dCas is aimed toward, this technique may inhibit both the transcription start and elongation phases (Bao *et al.* 2019). Another especially intriguing discovery was the potential to alter the degree of transcriptional repressed by weakening RNA/DNA connections by introducing mismatches in the gRNA/target link, which were brought about by mutations in the 5′ end of crRNA. The omega component of RNAP was fused with dCas to activate transcription. Following the recruitment and activation of the RNAP and an increase in gene transcription, dCas is guided to the target area (Feng *et al.* 2016). Single or multiplexed transcriptional control is made possible by the fusion of dCas or guide RNA to numerous activator or repressing sites, with varied degrees of modification, and specificity encouraged the fast appearance of novel Cas-based transcription modulators. Since the genomic DNA is unaltered, the changes in gene expression brought on by dCas are temporary due to its catalytically inactive nature. However, when dividing cells are the target, by attaching dCas to histone/DNA methyl transferase, demethylase, and acetyl transferase, changing DNA methylation or histone acetylation marks and triggering possibly inheritable genomic expression regulation, lasting alterations may be created (Sander & Joun, 2014).

There are less complex, DNA-independent techniques that are more effective than HDR, if the goal is to induce or fix a point mutation that just calls for a base substitution. The Cas nickase [Casn] produced by altering the RuvC effective site of Cas to aspartate-to-alanine would nick instead of DSB; the target DNA experiences single-stranded breaks. The anti-DNA strand cytosine [C] bases are delaminated by gRNA to create uracil [U], which has thymine base-pairing properties [T]; it is made possible by combining a cytidine deaminase enzyme with this Casn or dCas—the catalytically 'dead' variant of Cas. Intrinsic DNA replication and DNA repair converts the U base into a T base, results in a C–T [and G–A] substitution while creating a DSB. Contrast HDR-mediated Cas editing has been proven to increase the editing efficiency approximately upto 30 folds in numerous studies conducted in human cell lines with seamless integration of DNA, owing to the fact if HDR created desired changes, no indels should be observed. This technique was improved upon by later base editor generations to boost productivity and decrease off-target development.

Activation or inhibition of the transcription of bacterial genome showed how CRISPR-Cas may be used as a cutting-edge technique to regulate gene expression. As CRISPR research gained popularity, scientists switched from using bacteria to different cellular types. All types of cells and certain multicellular organisms, including human cell culture, animals, plants, yeast and a long list of others, would soon be the target of CRISPR-mediated modification.

APPLICATIONS OF CRISPR-CAS IN PLANTS

Crop Quality Improvement

Recently, crop quality has also been improved using a new gene-editing method known as the clustered regularly interspaced short palindromic repeats [CRISPR]-Cas technology. Due to its adaptability, it has emerged as the most common tool for crop modification. Due to its accuracy in precise gene editing, it has sped up crop breeding advances.

The market value of crops has mostly been determined by crop quality. Crop quality is often influenced by both external and internal factors. Size, color, texture and aroma are examples of physical and esthetically pleasing qualities that make up exterior qualitative traits. Contrarily, nutrients [such as protein, carbohydrates and fats] and bioactive substances [such as carotenoids, lycopene, amino butyric acid and flavonoids] are considered to be internal quality factors. Crop quality enhancement with CRISPR-Cas focuses on fruit esthetics, edible quality, fruit texture and nutritional value.

Biotic and Abiotic Tolerance in Plant

Insects drastically diminish yield via direct damage and also serve as disease vectors. Plant pathogens like bacteria, fungus and viruses produce many plant diseases that affect plants in different ways. The ability of plants to withstand various biotic stressors is being severely affected by the changing environment. It is estimated that biotic stressors cause direct crop losses of 20%–40%. The possibility for creating disease- and insect-resistant plants using the CRISPR-Cas system has been persuasively discussed in a number of papers. In addition to the biotic challenges, plants are also subject to a variety of abiotic stresses, such as temperature, salt, heavy metals and drought, all of which can impede crop development and result in large, sometimes catastrophic yield losses. Sensitivity genes are plant genes that intensify the harmful effects of abiotic stressors [*Se* genes]. Several plant species, including grain, vegetable and fruit crops, have benefited from genome-editing techniques, notably CRISPR-Cas, by altering these *Se* genes.

Improving Photosynthesis

Plant scientists have been working to increase crop photosynthesis for decades in order to increase production, including increasing flag leaf area, increasing CO_2 absorption, decreasing photorespiration and increasing flux through the Calvin cycle. CRISPR-based gene editing has also enhanced photosynthesis because of its crucial role in plant growth and development, grain and fruit yield and biomass production. For instance, CRISPR-based editing of the gene *NRP*1, a transcription factor-like gene that controls the expression of genes involved in photosynthesis, increased the efficiency, yield and biomass production of rice's photosynthetic process.

Yield Improvement

Plants have various grain yield properties. Quantitative trait loci [QTLs] that influence 1,000-grains weight [including seed size], quantity of grains per panicle, number of florets per panicle and number of panicles per plant all have an impact on grain production. Different QTLs that aid in increasing agricultural output have been discovered in various crops. 19 QTLs have been found in rice that regulate rice production.

By eliminating some undesirable genes or their transcription factor regulators, CRISPR-Cas has been used to enhance various aspects of plant architecture, including plant height, flag leaf size, leaf width, internodal distance, number of tillers, inflorescence structure, root growth and structure.

Improve Disease Resistance

Phytopathogens like viruses, bacteria and fungi pose a serious threat to crop productivity and are responsible for 30%–40% of crop losses worldwide. To feed the growing population, developing crop varieties that are resistant to disease is crucial. A number of technologies have been used to create crops that are resistant to disease. The genome-editing tool clustered regularly interspaced short palindromic repeat CRISPR-associated protein 9 [CRISPR-Cas] has transformed plant research, including other fields. Subsequent chapters of the book discuss the use of the CRISPR genome-editing tool to create plant disease resistance to bacteria, fungi and viruses. Host plants have been found to contain a number of disease susceptibility, recessive, negatively regulated genes and transcription factors that may play a part in a pathogen-friendly interaction. For the purpose of enhancing disease resistance, these genes were altered in a variety of plants, including rice, wheat, cucumber, cassava, cacao, grapes, citrus, apple and banana. In addition, applications and limitations in developing transgene free disease resistant crops will be discussed further.

FUTURE PROSPECTS

The use of the CRISPR-Cas system as a powerful genome-editing tool in recent years has greatly accelerated the progress of the biological sciences. In many archaeal and bacterial species, genome editing may benefit from more precise genetic modification. Additionally, type II CRISPR systems effectively induce desired changes in genome using single-RNP complexes, and also it is incriminated in research beyond genome-editing. CRISPR-Cas systems have successfully interfered in specific genomic regions after being converted as antibacterial systems (Li *et al.* 2016; Selle *et al.* 2015; Yosef *et al.* 2015). The effective use of this technique in other novel species has various obstacles that must be overcome. There is thus a pressing need to provide a standardized method for diseases to effectively transmit DNA (Shabbir *et al.* 2019). For genome-editing research to be effective, choosing the right tool for a particular DNA change as well as the delivery mechanism and kind of vectors is essential. Additionally, choosing the right CRISPR-Cas tool for a particular application is the key choice that will determine how successfully and efficiently genome editing is accomplished.

REFERENCES

Altpeter, F., Baisakh, N., Beachy, R., Bock, R., Capell, T., Christou, P., Dix, P. (2005). Particle bombardment and the genetic enhancement of crops: Myths and realities. *Molecular Breeding, 15*(3), 305–327.

Arora, L., & Narula, A. (2017). Gene editing and crop improvement using CRISPR-Cas9 system. *Frontiers in Plant Science, 8*, 1932.

Bao, A., Burritt, D.., Chen, H., Zhou, X., Cao, D., & Tran, L.-S. P. (2019). The CRISPR/Cas9 system and its applications in crop genome editing. *Critical Reviews in Biotechnology, 39*(3), 321–336.

Barrangou, R. (2013). CRISPR-Cas systems and RNA-guided interference. *Wiley Interdisciplinary Reviews: RNA, 4*(3), 267–278.

Barrangou, R., Fremaux, C., Deveau, H., Richards, M., Boyaval, P., Moineau, S., Horvath, P. (2007). CRISPR provides acquired resistance against viruses in prokaryotes. *Science, 315*(5819), 1709–1712.

Barrangou, R., & Marraffini, L. A. (2014). CRISPR-Cas systems: Prokaryotes upgrade to adaptive immunity. *Molecular Cell, 54*(2), 234–244.

Beerli, R. R., & Barbas, C. F. (2002). Engineering polydactyl zinc-finger transcription factors. *Nature Biotechnology, 20*(2), 135–141.

Belhaj, K., Chaparro-Garcia, A., Kamoun, S., Patron, N. & Nekrasov, V. (2015). Editing plant genomes with CRISPR/Cas9. *Current Opinion in Biotechnology, 32*, 76–84.

Bogdanove, A., Schornack, S., & Lahaye, T. (2010). TAL effectors: Finding plant genes for disease and defense. *Current Opinion in Plant Biology, 13*(4), 394–401.

Bolotin, A., Quinquis, B., Sorokin, A., & Ehrlich, S. D. (2005). Clustered regularly interspaced short palindrome repeats (CRISPRs) have spacers of extrachromosomal origin. *Microbiology, 151*(8), 2551–2561.

Bondy-Denomy, Garcia, B., Strum, S., Du, M., Rollins, M. F., Hidalgo-Reyes, Y., Davidson, A. R. (2015). Multiple mechanisms for CRISPR–Cas inhibition by anti-CRISPR proteins. *Nature, 526*(7571), 136–139.

Bortesi, L., & Fischer, R. (2015). The CRISPR/Cas9 system for plant genome editing and beyond. *Biotechnology Advances, 33*(1), 41–52.

Brouns, S. J., Jore, M. M., Lundgren, M., Westra, E. R., Slijkhuis, R. J., Snijders, A. P., Van Der Oost (2008). Small CRISPR RNAs guide antiviral defense in prokaryotes. *Science, 321*(5891), 960–964.

Carte, J., Christopher, R. T., Smith, J. T., Olson, S., Barrangou, R., Moineau, S., . . . Terns, M. P. (2014). The three major types of CRISPR-Cas systems function independently in CRISPR RNA biogenesis in S treptococcus thermophilus. *Molecular Microbiology, 93*(1), 98–112.

Carte, J., Wang, R., Li, H., Terns, R. M., & Terns, M. P. (2008). Cas6 is an endoribonuclease that generates guide RNAs for invader defense in prokaryotes. *Genes & Development, 22*(24), 3489–3496.

Černý, J., & Stříž, I. (2019). Adaptive innate immunity or innate adaptive immunity? *Clinical Science, 133*(14), 1549–1565.

Charpentier, E., & Marraffini, L. A. (2014). Harnessing CRISPR-Cas9 immunity for genetic engineering. *Current Opinion in Microbiology, 19*, 114–119.

Chen, J. S., Dagdas, Y. S., Kleinstiver, B. P., Welch, M. M., Sousa, A. A., Harrington, L. B., . . . Doudna, J. A. (2017). Enhanced proofreading governs CRISPR–Cas9 targeting accuracy. *Nature, 550*(7676), 407–410.

Chen, W., Zhang, Y., Yeo, W.-S., Bae, T., & Ji, Q. (2017). Rapid and efficient genome editing in Staphylococcus aureus by using an engineered CRISPR/Cas9 system. *Journal of the American Chemical Society, 139*(10), 3790–3795.

Chevallereau, A., Meaden, S., van Houte, S., Westra, E. R., & Rollie, C. (2019). The effect of bacterial mutation rate on the evolution of CRISPR-Cas adaptive immunity. *Philosophical Transactions of the Royal Society B, 374*(1772), 20180094.

Collias, D., & Beisel, C. L. (2021). CRISPR technologies and the search for the PAM-free nuclease. *Nature Communications, 12*(1), 1–12.

Cong, L., Ran, F. A., Cox, D., Lin, S., Barretto, R., Habib, N. Marraffini, L. A. (2013). Multiplex genome engineering using CRISPR/Cas systems. *Science, 339*(6121), 819–823.

Cox, D. B., Gootenberg, J. S., Abudayyeh, O. O., Franklin, B., Kellner, M. J., Joung, J., & Zhang, F. (2017). RNA editing with CRISPR-Cas13. *Science, 358*(6366), 1019–1027.

Deretic, V. (2005). Autophagy in innate and adaptive immunity. *Trends in Immunology, 26*(10), 523–528.

Eeckhaut, T., Lakshmanan, P. S., Deryckere, D., Van Bockstaele, E., & Van Huylenbroeck, J. (2013). Progress in plant protoplast research. *Planta, 238*(6), 991–1003.

Fauser, F., Schiml, S., & Puchta, H. (2014). Both CRISPR/C as-based nucleases and nickases can be used efficiently for genome engineering in A rabidopsis thaliana. *The Plant Journal, 79*(2), 348–359.

Feng, Z., Zhang, B., Ding, W., Liu, X., Yang, D.-L., Wei, P., Mao, Y. (2013). Efficient genome editing in plants using a CRISPR/Cas system. *Cell Research, 23*(10), 1229–1232.

Fichtner, F., Urrea Castellanos, R., & Ülker, B. (2014). Precision genetic modifications: A new era in molecular biology and crop improvement. *Planta, 239*(4), 921–939.

Fonfara, I., Le Rhun, A., Chylinski, K., Makarova, K. S., Lecrivain, A.-L., Bzdrenga, J., Charpentier, E. (2014). Phylogeny of Cas9 determines functional exchangeability of dual-RNA and Cas9 among orthologous type II CRISPR-Cas systems. *Nucleic Acids Research, 42*(4), 2577–2590.

Friedland, A. E., Tzur, Y. B., Esvelt, K. M., Colaiácovo, M. P., Church, G. M., & Calarco, J. A. (2013). Heritable genome editing in C. elegans via a CRISPR-Cas9 system. *Nature Methods, 10*(8), 741–743.

Gao, W., Long, L., Tian, X., Xu, F., Liu, J., Singh, P. K., Song, C. (2017). Genome editing in cotton with the CRISPR/Cas9 system. *Frontiers in Plant Science, 8*, 1364.

Gilbert, L. A., Horlbeck, M. A., Adamson, B., Villalta, J. E., Chen, Y., Whitehead, E. H.,Bassik, M. C. (2014). Genome-scale CRISPR-mediated control of gene repression and activation. *Cell, 159*(3), 647–661.

Gilbert, L. A., Larson, M. H., Morsut, L., Liu, Z., Brar, G. A., Torres, S. E., Doudna, J. A. (2013). CRISPR-mediated modular RNA-guided regulation of transcription in eukaryotes. *Cell, 154*(2), 442–451.

Hayes, R. P., Xiao, Y., Ding, F., Van Erp, P. B., Rajashankar, K., Bailey, S., Ke, A. (2016). Structural basis for promiscuous PAM recognition in type I–E Cascade from E. coli. *Nature, 530*(7591), 499–503.

Hille, F., Richter, H., Wong, S. P., Bratovič, M., Ressel, S., & Charpentier, E. (2018). The biology of CRISPR-Cas: Backward and forward. *Cell, 172*(6), 1239–1259.

Hryhorowicz, M., Lipiński, D., Zeyland, J., & Słomski, R. (2017). CRISPR/Cas9 immune system as a tool for genome engineering. *Archivum immunologiae et therapiae experimentalis, 65*(3), 233–240.

Ishino, Y., Krupovic, M., & Forterre, P. (2018). History of CRISPR-Cas from encounter with a mysterious repeated sequence to genome editing technology. *Journal of Bacteriology, 200*(7), e00580–00517.

Ishino, Y., Shinagawa, H., Makino, K., Amemura, M., & Nakata, A. (1987). Nucleotide sequence of the iap gene, responsible for alkaline phosphatase isozyme conversion in Escherichia coli, and identification of the gene product. *Journal of Bacteriology, 169*(12), 5429–5433.

Jasin, M., & Rothstein, R. (2013). Repair of strand breaks by homologous recombination. *Cold Spring Harbor Perspectives in Biology, 5*(11), a012740.

Jinek, M., Chylinski, K., Fonfara, I., Hauer, M., Doudna, J. A., & Charpentier, E. (2012). A programmable dual-RNA–guided DNA endonuclease in adaptive bacterial immunity. *Science, 337*(6096), 816–821.

Kearns, N. A., Genga, R. M., Enuameh, M. S., Garber, M., Wolfe, S. A., & Maehr, R. (2014). Cas9 effector-mediated regulation of transcription and differentiation in human pluripotent stem cells. *Development, 141*(1), 219–223.

Kim, H., & Kim, J.-S. (2014). A guide to genome engineering with programmable nucleases. *Nature Reviews Genetics, 15*(5), 321–334.

Koonin, E. V., & Makarova, K. S. (2009). CRISPR-Cas: An adaptive immunity system in prokaryotes. *F1000 Biology Reports, 1*.

Koonin, E. V., & Makarova, K. S. (2019). Origins and evolution of CRISPR-Cas systems. *Philosophical Transactions of the Royal Society B, 374*(1772), 20180087.

Koonin, E. V., Makarova, K. S., & Zhang, F. (2017). Diversity, classification and evolution of CRISPR-Cas systems. *Current Opinion in Microbiology, 37*, 67–78.

Li, J.-F., Norville, J. E., Aach, J., McCormack, M., Zhang, D., Bush, J., Sheen, J. (2013). Multiplex and homologous recombination–mediated genome editing in Arabidopsis and Nicotiana benthamiana using guide RNA and Cas9. *Nature Biotechnology, 31*(8), 688–691.

Li, J., Shen, S., Chen, J., Liu, W., Li, X., Zhu, Q., Wang, M. (2018). Accurate annotation of accessible chromatin in mouse and human primordial germ cells. *Cell Research, 28*(11), 1077–1089.

Li, R., Fang, L., Tan, S., Yu, M., Li, X., He, S., Wu, M. (2016). Type I CRISPR-Cas targets endogenous genes and regulates virulence to evade mammalian host immunity. *Cell Research, 26*(12), 1273–1287.

Liu, K. I., Ramli, M. N. B., Woo, C. W. A., Wang, Y., Zhao, T., Zhang, X., Chua, M. Z. H. (2016). A chemical-inducible CRISPR–Cas9 system for rapid control of genome editing. *Nature Chemical Biology, 12*(11), 980–987.

Makarova, K. S., Haft, D. H., Barrangou, R., Brouns, S. J., Charpentier, E., Horvath, P., Yakunin, A. F. (2011). Evolution and classification of the CRISPR–Cas systems. *Nature Reviews Microbiology, 9*(6), 467–477.

Makarova, K. S., Wolf, Y. I., Alkhnbashi, O. S., Costa, F., Shah, S. A., Saunders, S. J., Haft, D. H. (2015). An updated evolutionary classification of CRISPR–Cas systems. *Nature Reviews Microbiology, 13*(11), 722–736.

Marraffini, L. A. (2015). CRISPR-Cas immunity in prokaryotes. *Nature, 526*(7571), 55–61.

Marraffini, L. A., & Sontheimer, E. J. (2010). CRISPR interference: RNA-directed adaptive immunity in bacteria and archaea. *Nature Reviews Genetics, 11*(3), 181–190.

Mohanraju, P., Makarova, K. S., Zetsche, B., Zhang, F., Koonin, E. V., & Van der Oost, J. (2016). Diverse evolutionary roots and mechanistic variations of the CRISPR-Cas systems. *Science, 353*(6299), aad5147.

Mojica, F. J., & Rodriguez-Valera, F. (2016). The discovery of CRISPR in archaea and bacteria. *The FEBS Journal, 283*(17), 3162–3169.

Morange, M. (2015). What history tells us XXXVII. CRISPR-Cas: The discovery of an immune system in prokaryotes. *Journal of Biosciences, 40*(2), 221–223.

Nekrasov, V., Staskawicz, B., Weigel, D., Jones, J. D., & Kamoun, S. (2013). Targeted mutagenesis in the model plant Nicotiana benthamiana using Cas9 RNA-guided endonuclease. *Nature Biotechnology, 31*(8), 691–693.

Nielsen, A. A., & Voigt, C. A. (2014). Multi-input CRISPR/C as genetic circuits that interface host regulatory networks. *Molecular Systems Biology, 10*(11), 763.

Nødvig, C. S., Nielsen, J. B., Kogle, M. E., & Mortensen, U. H. (2015). A CRISPR-Cas9 system for genetic engineering of filamentous fungi. *PloS One, 10*(7), e0133085.

Pan, S., Li, Q., Deng, L., Jiang, S., Jin, X., Peng, N., Li, Y. (2019). A seed motif for target RNA capture enables efficient immune defence by a type III-B CRISPR-Cas system. *RNA Biology, 16*(9), 1166–1178.

Parsi, K. M., Hennessy, E., Kearns, N., & Maehr, R. (2017). Using an inducible CRISPR-dCas9-KRAB effector system to dissect transcriptional regulation in human embryonic stem cells. In *Eukaryotic Transcriptional and Post-Transcriptional Gene Expression Regulation* (pp. 221–233): Springer.

Patterson, A. G., Yevstigneyeva, M. S., & Fineran, P. C. (2017). Regulation of CRISPR–Cas adaptive immune systems. *Current Opinion in Microbiology, 37*, 1–7.

Peng, R., Lin, G., & Li, J. (2016). Potential pitfalls of CRISPR/Cas9-mediated genome editing. *The FEBS Journal, 283*(7), 1218–1231.

Piatek, A., Ali, Z., Baazim, H., Li, L., Abulfaraj, A., Al-Shareef, S., Mahfouz, M. M. (2015). RNA-guided transcriptional regulation in planta via synthetic dC as9-based transcription factors. *Plant Biotechnology Journal, 13*(4), 578–589.

Ran, F., Hsu, P. D., Wright, J., Agarwala, V., Scott, D. A., & Zhang, F. (2013). Genome engineering using the CRISPR-Cas9 system. *Nature Protocols, 8*(11), 2281–2308.

Sampson, T. R., & Weiss, D. S. (2014). Exploiting CRISPR/C as systems for biotechnology. *Bioessays, 36*(1), 34–38.

Sander, J. D., & Joung, J. K. (2014). CRISPR-Cas systems for genome editing, regulation and targeting. *Nature Biotechnology, 32*(4), 347.

Schindele, P., Wolter, F., & Puchta, H. (2018). Transforming plant biology and breeding with CRISPR/Cas9, Cas12 and Cas13. *FEBS Letters, 592*(12), 1954–1967.

Selle, K., & Barrangou, R. (2015). Harnessing CRISPR–Cas systems for bacterial genome editing. *Trends in Microbiology, 23*(4), 225–232.

Shabbir, M. A. B., Shabbir, M. Z., Wu, Q., Mahmood, S., Sajid, A., Maan, M. K., Yuan, Z. (2019). CRISPR-cas system: Biological function in microbes and its use to treat antimicrobial resistant pathogens. *Annals of Clinical Microbiology and Antimicrobials, 18*(1), 1–9.

Shan, Q., Wang, Y., Li, J., Zhang, Y., Chen, K., Liang, Z., Qiu, J.-L. (2013). Targeted genome modification of crop plants using a CRISPR-Cas system. *Nature Biotechnology, 31*(8), 686–688.

Shmakov, S., Abudayyeh, O. O., Makarova, K. S., Wolf, Y. I., Gootenberg, J. S., Semenova, E., . . . Severinov, K. (2015). Discovery and functional characterization of diverse class 2 CRISPR-Cas systems. *Molecular Cell, 60*(3), 385–397.

Tadić, V., Josipović, G., Zoldoš, V., & Vojta, A. (2019). CRISPR/Cas9-based epigenome editing: An overview of dCas9-based tools with special emphasis on off-target activity. *Methods, 164*, 109–119.

Terns, M. P., & Terns, R. M. (2011). CRISPR-based adaptive immune systems. *Current Opinion in Microbiology, 14*(3), 321–327.

Terns, R. M., & Terns, M. P. (2014). CRISPR-based technologies: Prokaryotic defense weapons repurposed. *Trends in Genetics, 30*(3), 111–118.

Tyagi, S., Kumar, R., Das, A., Won, S. Y., & Shukla, P. (2020). CRISPR-Cas9 system: A genome-editing tool with endless possibilities. *Journal of Biotechnology, 319*, 36–53.

Van der Oost, J., Jore, M. M., Westra, E. R., Lundgren, M., & Brouns, S. J. (2009). CRISPR-based adaptive and heritable immunity in prokaryotes. *Trends in Biochemical Sciences, 34*(8), 401–407.

Van Der Oost, J., Westra, E. R., Jackson, R. N., & Wiedenheft, B. (2014). Unravelling the structural and mechanistic basis of CRISPR–Cas systems. *Nature Reviews Microbiology, 12*(7), 479–492.

Vejnar, C. E., Moreno-Mateos, M. A., Cifuentes, D., Bazzini, A. A., & Giraldez, A. J. (2016). Optimized CRISPR–Cas9 system for genome editing in zebrafish. *Cold Spring Harbor Protocols, 2016*(10), pdb. prot086850.

Wang, H., Yang, H., Shivalila, C. S., Dawlaty, M. M., Cheng, A. W., Zhang, F., & Jaenisch, R. (2013). One-step generation of mice carrying mutations in multiple genes by CRISPR/Cas-mediated genome engineering. *Cell, 153*(4), 910–918.

Wang, Y., Zhang, Z.-T., Seo, S.-O., Lynn, P., Lu, T., Jin, Y.-S., & Blaschek, H. P. (2016). Bacterial genome editing with CRISPR-Cas9: Deletion, integration, single nucleotide modification, and desirable "clean" mutant selection in Clostridium beijerinckii as an example. *ACS Synthetic Biology, 5*(7), 721–732.

Westra, E. R., Buckling, A., & Fineran, P. C. (2014). CRISPR–Cas systems: Beyond adaptive immunity. *Nature Reviews Microbiology, 12*(5), 317–326.

Wiedenheft, B., Sternberg, S. H., & Doudna, J. A. (2012). RNA-guided genetic silencing systems in bacteria and archaea. *Nature, 482*(7385), 331–338.

Xie, K., & Yang, Y. (2013). RNA-guided genome editing in plants using a CRISPR–Cas system. *Molecular Plant, 6*(6), 1975–1983.

Xue, C., Seetharam, A. S., Musharova, O., Severinov, K., J. Brouns, S. J., Severin, A. J., & Sashital, D. G. (2015). CRISPR interference and priming varies with individual spacer sequences. *Nucleic Acids Research, 43*(22), 10831–10847.

Yamano, T., Zetsche, B., Ishitani, R., Zhang, F., Nishimasu, H., & Nureki, O. (2017). Structural basis for the canonical and non-canonical PAM recognition by CRISPR-Cpf1. *Molecular Cell, 67*(4), 633–645. e633.

Yang, H., Gao, P., Rajashankar, K. R., & Patel, D. J. (2016). PAM-dependent target DNA recognition and cleavage by C2c1 CRISPR-Cas endonuclease. *Cell, 167*(7), 1814–1828. e1812.

Yosef, I., Manor, M., Kiro, R., & Qimron, U. (2015). Temperate and lytic bacteriophages programmed to sensitize and kill antibiotic-resistant bacteria. *Proceedings of the National Academy of Sciences, 112*(23), 7267–7272.

You, L., Ma, J., Wang, J., Artamonova, D., Wang, M., Liu, L.Wang, Y. (2019). Structure studies of the CRISPR-Csm complex reveal mechanism of co-transcriptional interference. *Cell, 176*(1–2), 239–253. e216.

2 Plant Defense and Disease Resistance Mechanisms

INTRODUCTION

Plant pathogens cause diseases in diverse plant parts using various pathogenicity mechanims, which causes significant monetary loss. Via the recognition of plant surface neurotransmitters by molecular processes, the production of pathogenicity and effector proteins, and the restriction of plant defense limit bacteria to infect plants effectively (Khan *et al.* 2022). Microbes evolve new characters in order to continue attacking plants, and plants evolve defenses to interact with and accommodate such new microbial varieties. When a plant is affected by a pathogen, signal routes are set off, which causes the interpretation of defense genes in the plant, activating defense responses against the microorganism (Wei *et al.* 2022). The method of microbial pathogens is restricted, and its gene of expression is repressed as quickly as the immune system of the plant initiates this response. Both internal elicitors (structure of plant) and external elicitors (biotic or abiotic) can trigger these defense responses. Endogenous substances are now frequently used in agricultural production throughout the world to reduce losses by various pathogens.

Along with other traditional strategies, using genetic engineering could be effective for developing resistance against diseases (Deng *et al.* 2022). In comparison, the use of biotic stimulants to prevent infection in plants in environmental circumstances could cause shifts in the framework of initiation of resistance and the structure of the defense signals due to the intricate interaction that takes place between microbial, plant, biological activator constituents and the surroundings. For optimum disease management, it is necessary to manage the interconnections between such three microbial agents, plant pathogenic organisms and biological causative agents along with the atmosphere (D'Andrea & Dullerud, 2003). Various environmental factors can cause changes in plants, pathogenic organisms and biological elements, which could impact the resistance process.

Apparently, a less effective biological management is accomplished in the field compared to that accomplished in the controlled situations of the glasshouse and research center as a result of the complicated quadruple impacts (microbial plant activator agents' surroundings) that eventuate in between variables (Fosha *et al.* 1970). Numerous studies have focused on the contribution of abiotic agents in the initiation of resistance against phytopathogens. One illustration is gamma-aminobutyric acid (GABA) that has been effectively used in agricultural practice for defense priming in various crops. It has now been shown that a wide variety of

DOI: 10.1201/b22901-2

organic substances, including oligosaccharides, glycosides, amides, micronutrients, carboxylic acids and aromatic compounds can improve the defense priming reaction in plants (Sandell *et al.* 1975). Hexanoic acid, a straightforward substance, demonstrates a powerful natural stimulatory capacity to shield plants from a variety of diseases by causing callose accumulation and establishing salicylic acid (SA) and jasmonic acid (JA) mechanisms (Sandell *et al.* 1975). Organic polysaccharide hydrolysis yields oligosaccharide goods, which can act as inducers to trigger resilience and modify the utterance of plant defense genes. *Agrobacterium* spp. fermentation content oligocurdlan, which was recently shown to demonstrate stimulate systemic resistance against *Phytophthora infestans* in potatoes, is an illustration of how microorganism's items can also stimulate defense responses in plants. Other instances have included the oligosaccharides that are naturally found in green and brown cyanobacteria and can trigger plant defense signals. In the agricultural and horticultural industries, these substances are also utilized as biofertilizers (Rothman & Wagner, 2003).

Today, a variety of commercial products for plant protection that usually contain oligosaccharides are heavily promoted. In order to control microbial maladies in the research area, bioactive antioxidants like oligosaccharides are now showing promise as alternatives to synthetic fungicides. A good substitute method for using chemical pesticides to regulate diseases in plants is stimulated resistance (Shayeghi *et al.* 2009). A major problem is discovering new organic inoculant causes and examining how they affect plant defense. The primary defense mechanisms used by crops to resist microbial attacks are described, with a particular emphasis on the roles played by ALG (Alginate) and AOS (Alginate oligosaccharides) in the initiation of resistance against plant diseases (Venkat *et al.* 2008).

DISEASES IN PLANTS

Fungal Diseases in Plants

Fungi, which may also proliferate either sexually or asexually, are responsible for around 80% of crop maladies. In fact, they were able to invade new areas of the plants or even consume new organelles. Air, water, soil and animals all have a role in dispersing fungal spores, which may then infect and kill other seedlings (Smith *et al.* 2007). Warm, damp places are ideal for mold growth. Fungi may not only cause devastating diseases, including anthracnose, late blight, apple scab, club roots, black spot, damping off and powdery mildew, but they also play a crucial part in plant growth, especially when they form mycorrhizal associations and get entangled with plant root systems. Fungi-like pathogens (FLPs) are able to take responsibility for more plant diseases compared to any other classification, despite the fact that there are an estimated 8,000 various types of plant-like pests (Kevan *et al.* 1975). Microbes that cause downy mildew plant diseases like *Phytophthora* and *Pythium* are now understood to connect to a different taxonomic group.

Viruses and Viroids

Viruses and viroids are the types of pathogens that are the most challenging to manage, particularly plant-infecting pathogens. In most cases, destroying diseased plants is the most effective method of disease management. This is because, the chemical compounds designed to inhibit plant viruses and the viral vectors that transmit them have not been proved to be successful. There have been around 300 different plant viruses, but due to the fact that these germs are able to evolve, there are always new variations appearing (Jeandet *et al.* 2022). Most phages are either rod-shaped (isometric) or polyhedral (polyhedral). Deformities such as leaf rolls and abnormally narrow leaf development, as well as a general yellowing, are symptoms of viral infection (Bozbuga *et al.* 2022).

Plant Diseases Caused by Bacteria

The rod-shaped bacteria cause many diseases in plants. The stomata in the leaves and any pores in the plant's cells are natural entry points for bacteria. When pathogens invade a plant, they travel through its vascular system (its water and nutrient channels) and eventually cause the plant to wilt (Sharma *et al.* 2022b). Plant tissue breakdown and swelling are also common indicators of bacterial infection. Bacteria may spread in a number of ways, including via the air, insects, infected objects and soil. T. J. Burrill (1877–1855) of the University of Illinois first identified the bacterium, *Erwinia amylovora*, responsible for pear and apple fire blight (Golea & Hideg, 2022). The most prevalent harmful bacteria found in herbs include *Erwinia, Pectobacterium, Pantea, Pseudomonas, Ralstonia, Burkholderia, Acidovorax, Xanthomonas, Clavibacter, Streptomyces, Xylella, Spiroplasma* and *Phytoplasma*. Numerous symptoms caused by plant bacterial infections include galls and overgrowths, wilts, leaf stains, soft rots, scabs and cankers. Walled microorganisms multiply inside the gaps between the cells as opposed to viral infections, which remain within the host tissue (Sharma *et al.* 2022a).

Nematodes Causing Plant Diseases

Nematodes are worm-like creatures that feed on micronutrients. They develop through four different stages before they become adults and produce eggs and eventually turn into larvae. Nematodes are only alive for 30 days, but they have a dormant lifespan of over 30 years (Mendoza-de Gives, 2022). Nematicides are organic compounds that are used to manage nematode infections. Because of their resistance to nematodes, marigolds are frequently cultivated to keep soil from becoming infected. Nematodes primarily target root growth; however, they are also capable of externally or internally destroying other plant parts. They target a variety of plant species, such as maize, rye and onions, and they thrive in hot, sandy and humid soil. However, not all nematodes have had any impact on plants. In fact, some of them have been employed in the management of other pests, including armyworms, caterpillars and beetle grubs (Pereira & González-Solís, 2022).

Plant Immunity against Diseases

Pathogen or microbe-associated molecular patterns (PAMPs or MAMPs), also known as evolutionarily conserved microorganisms signal molecules, can be detected by plant receptors that recognize patterns. The PAMP compounds serve as an effective form that plants use to recognize the presence of pathogenic organisms, because they are crucial for microbial survival (Sanguankiattichai *et al.* 2022). PAMP-triggered immunity, which offers defense against nonhost pathogens and restricts diseases caused by virulent pathogens, is brought on when plant pattern recognition neurotransmitters recognize PAMPs (Ramírez-Zavaleta *et al.* 2022).

However, pathogens have developed mechanisms to suppress plant defenses brought on by pathogenicity signals and genes, as well as adaptations to their host plants. In return, the plant's hypersensitive response (HR) and systemic acquired resilience (SAR) develops, providing the plant with disease resistance termed as effector-triggered response (Soto *et al.* 2022). These R proteins are either involved directly or indirectly in diagnosing the pathogen's effector proteins. Pathogen enzymatic activity often releases components of plant cell walls and cutin, both of which may activate immunological reactions in plants. The promoter defense and pathogen-associated molecular pattern (PAMP)-triggered immunity pathways activate signal transduction and transcription factors, which in turn suppress the growth of pathogens and the manifestation of disease symptoms. The accumulation of defense hormones, such as SA, ET and JA, and the production of reactive oxygen species (ROS) are also detected (Chiu *et al.* 2022). A crucial regulatory mechanism for plant immunity has also been identified as a crosstalk between the SA and JA–ET signaling pathways. Although defense genes are present in plants, they are frequently latent in healthy environments. However, a practice known as mediated resistance allows for the induction of these defense genes in plants using any inducer (Min *et al.* 2022). The activator activates the plant's defense mechanism in response to successive invading pathogens, preventing the spread of disease. Wide-ranging defense mechanisms are activated by mediated resistance, and the defense indicators in this mechanism result in both SAR and induced systemic resistance (ISR) (Soto *et al.* 2022).

SYSTEMIC ACQUIRED RESISTANCE (SAR)

SAR is a kind of plant defense response that provides constant protection against many different plant diseases. These molecules are responsible for depending on the systemic signal that is caused by the interaction between the plants and the pathogens. Substance P (SA), lipid-based signal molecules and reactive oxygen species (ROS) are all components of the systemic transmissions linked to SAR. The production of SA—a signaling chemical and the accumulation of PR proteins are both linked to SAR (Aboulila & Pathology, 2022). A large variety of necrosis- and hypersensitive-response-causing microorganisms may certainly trigger SAR in several plant species. This kind of resilience is both durable and effective against such a wide range of infections.

The upregulation for isochorismate synthase 1 (ICS1), which is accountable for one of the important enzymes in the system that makes SA, is increased in response to viral pathogens, SA is a defense hormone (Gao *et al.* 2022). Improved articulating of palmitic acid and its metabolites has boost another defensive signal for SAR in the primed guard cells of *arabidopsis* plants. SAR relies on a loop formed by NO and ROS—the first chemical signals in systemic immunity. Also, reactive oxygen species (ROS) can accelerate the chemical breakdown of unsaturated fats, which is necessary for inducing SAR in plants. SA binds to the H_2O_2 enzymes CAT and ascorbate peroxidase during SAR, blocking their activity and increasing H_2O_2 levels. This increase triggers signal transduction that ultimately induces genes involved in pathogenicity and pathogen reduction (Jiang *et al.* 2022).

INDUCED SYSTEMIC RESISTANCE (ISR)

Plant growth promoting rhizobacteria (PGPR) colonizes the surface of the roots, blocking microbial entry while enhancing systemic resistance in the plant. For the initiation of ISR, a particular confirmation response between the plant and the rhizobacteria is required. ISR can be induced by rhizobacterial factors like siderophores, lipopolysaccharides, antibiotics, quorum-sensing molecules and flagellated proteins. The plant's defenses against additional invading forces are boosted when this kind of resistance occurs (Yu *et al.* 2022).

ISR is a nonspecific reaction, as evidenced by its generalized activity against various pathogens. JA and ET are play main role in a pathway that typically activates ISR. Though, several PGPR has been reported to activate SA-dependent mechanisms, beneficial rhizobacteria frequently activate JA–ET-dependent pathways (Nguyen *et al.* 2022). For instance, NPR1, a common regulator of the SAR and ISR pathways, controls the transcription of genes related to SA-responsive pathogenesis (Thankappan *et al.* 2022).

Abiotic Inducers of Disease Resistance in Plants

Compounds that act at different points along the signal transduction pathways responsible for disease impedance as well as defense against abiotic and biotic stresses are referred to as abiotic stimulants. Two substances—2, 6-dichloroisonicotinic acid and its methyl ester—were the first synthesized substances which show prime defense responses within plants (Riseh *et al.* 2022). These chemical causative agents intensify a variety of cellular reactions, such as modifications in ion transport across the plasma membrane, production of antibacterial bioactive molecules (such as phytoalexins, cell wall phenolic and lignin-like polymers) and activation of defense genes. Chemical elicitors cause resistance that is long lasting as well as broad spectrum, and many of them offer disease control ranging from 20% to 85%. For detail, exposing plants to substances like SA, probenazole, benzothiadiazole and gamma-aminobutyric acid can all result in the development of resistance (Lazazzara *et al.* 2022). For example, plants exposed to

SA, probenazole, benzothiadiazole and gamma-aminobutyric acid may develop resistance to a wide range of pathogens. After treating tomato seeds with gamma-aminobutyric acid or JA, it was found that the mediated resistance reliable and was dependent on the induction of gene regulation. In a similar way, acibenzolar-S-methyl treatment of faba beans led to SAR against rust and ascochyta blight diseases in both greenhouse and field conditions, and this protection persisted over many weeks after acibenzolar-S-methyl application. ALG is a particularly ecofriendly polymer compound that works well with other compounds to activate plant immune defenses (Ebrahimi-Zarandi & Skorik, 2022). By employing these abiotic components, chemical-based pesticide agents' negative impacts on humans and other non-target organisms are avoided. These substances can make plants more resistant to pathogens and boost the expression of SA-dependent defense mechanisms. The benefits of polysaccharides and the methods by which ALG induces resistance against plant pathogens are covered in the sections that follow (Swain *et al.* 2022).

Polysaccharides as Plant Defense Inducers

The plant defense mechanism is crucial to its capacity to resist pathogens, making it a good target for disease control research. PAMPs with structural or chemical similarities to their pathogens are used by plants to identify pathogens. As a result, oligosaccharides with structures resembling those of pathogen cellular membranes or other frameworks can also function as PAMPs to stimulate the plant immune response (Shokrollahi *et al.* 2022). To lessen the environmental impacts of today's agricultural chemicals, ecofriendly alternatives must be promoted. The utilization of plant extracts and oils, microbial (bacteria, fungi and microalgae) extracts, seaweed extracts and polysaccharides can induce defense resistance, according to numerous reports from the recent decades (Aitouguinane *et al.* 2022).

With excellent structural complexity and biological activity, polysaccharides are now the best and most ecologically responsible biological supplies for developing plant pathogen internal resistance. On the biochemical and metabolomics indicators connected to defense pathways in tomato plants, the impacts of polysaccharides derived from microalgae and cyanobacteria were assessed (Oladosu *et al.* 2022). The phenylalanine ammonia-lyase, chitinase, -1,3-glucanase and peroxidase actions of the enzymes in tomato leaves were enhanced by the polysaccharides derived from *Phaeodactylum triocnutum*, *Desmodesmus* sp. and *Porphyridium* sp. Additionally, GC-MS metabolomics research showed that tomato leaves' metabolite contents, including those for fatty acids, alkanes and phytosterols were changed by polysaccharides (Yadav *et al.* 2022). Low-molecular-weight carbohydrates called oligosaccharides are produced when polysaccharides break down. Numerous living things are biologically active when these substances are present. They control particular plant growth mechanisms of plants, including cell morphogenesis and the pH-based development of flowers or calluses. Through the use of oligosaccharides, soil fertility can be improved, and plant defense mechanisms against biotic and abiotic stresses can be triggered (Ezati *et al.* 2022).

MECHANISMS OF DISEASE RESISTANCE IN PLANTS

Genetics of Disease Resistance

By traversing a resilient parent with an immune-compromised parent and analyzing the F2 single plants, lines in greater selfing generations or doubled-haploid populaces, the genetics of resistor can be studied. It is crucial for the breeder to distinguish between resistors that segregate quantitatively, i.e., reveal a constant slope evaluated as disease quantitatively or qualitatively, i.e., produce distinct groups evaluated as infectious disease type (Wulff & Krattinger, 2022). Quantitative characteristics are commonly referred to analyses of variance, whereas qualitative characteristics are typically tested using chi-squared tests of anticipated Mendelian segregation proportions (ANOVA). Moreover, this classification only relates to studies where the genetic influence (the ratio of genotypic to phenotypic traits) is significant, because a monogenic characteristic can also produce a constant phenotypic allocation when the impact of mistake and/or external conditions is substantial. Disease-resistant heritability is a complicated thing that involves molecular genetics, quantitative genetic factors and plant breeding techniques (Zhao *et al.* 2022).

Qualitative resistance to disease is defined as reactance variability caused by allelic mutation at only one or two R genes (resistance genes), with enough allele impacts to reliably determine an individual's genotype for dissent from its trait at single plant level, regardless of the variability (Riseh *et al.* 2022). Quantitative disease resilience (QDR) was also defined in a number of ways, including biologically as resistance relying on the concerted action of numerous modestly effective genes and phenotypically as the decrease but not complete removal of diseases linked to the most vulnerable phenotypes. Quantitative trait loci (QTL) are loci linked to variability in quantitative characters. Most QDR is, incomplete and has a multigenic foundation. Resistance breeding typically aims to develop long-lasting disease resistance, which is defined as resistance that is effective over a wide area for an extended period of time. Despite being challenging to test, QDR is thought to offer more long-lasting disease resistance than qualitative resistance (Webster & He, 2022).

Qualitative Resistances

Hypersensitivity (HR) generates monogenic resistances. When the pathogen's gene product, encrypted by the avirulence (*Avr*1) gene, is recognized by the plant's target gene, encrypted by the related resistance genes, an irreconcilable response follows, destroying the invading host cells. If the plant only has vulnerable alleles at this region, the response is always compatible (susceptible). If the pathogen is R1 (avrl) virulent, all responses are compatible (Sett *et al.* 2022). According to the gene-for-gene theory, each plant resistance gene has a matching pathogen infectivity gene. Flor was the first to simultaneously analyze pathogen infectivity and flax resistance. This concept has been validated by multiple plant–pathogen links with qualitative resistance inheritance. Because one dissenting allele is enough to induce resistance, the technique may be simplified to yield the quadratic test if the resistance gene is mostly inherited (Sun *et al.* 2022).

Most fungal and viral resistance genes (*R* genes) include leucine-rich repeats (NB-LRR). Fast oxidant production indicates HR. *R* and *Avr* genes are inherited primarily; however, recessive *R* genes may also cause viral resistance. Seedling or leaf-segment testing measures all-stage resistances. Based on the gene-for-gene concept, race selectivity is a strong predictor of qualitative resistances. Many microorganisms react quickly to R genes. Mutations in the Avr gene result in the emergence of new, virulent pathotypes (Chow *et al.* 2022). Viruses readily infect resistant cultivars, and by evolving to virulence, the virus rendered the R gene ineffective, thereby destroying resistance. Noxious pathotypes may spread swiftly in bacterial populations if the mutation doesn't affect the pathogen's aerobic ability. Numerous bacteria, fungi, nematodes, viruses and insects interact with plants gene-for-gene. Most organisms are biotrophs, such as rusts (*Puccinia* spp.), powdery mildew (*Blumeria graminis*), smuts (*Ustilago* spp.), and bunts (*Tilletia* spp.) and potato blight (*P. infestans*), but others are necrotrophs, such as rice blast (*Magnaporthe grisea*) and *septoria tritici* (*Setosphaeria turcica*) (Osuna-Caballero *et al.* 2022). Most *R* genes are active throughout a plant's lifespan; therefore, seedling resistance may be evaluated. A simple glasshouse or lab test utilizing a seedling or leaf segment may determine this. Cultivars with just racial-specific genes may disappear in a few decades and losses arise. Each pathosystem contains several *R* genes. 58 genes for stem rust (Sr) caused by *Puccinia graminis*, 70 genes for leaf rust (Lr) driven by *Puccinia triticina*, and at least 53 specifically and 39 briefly assigned genes for yellow rust (Yr) triggered by *Puccinia striiformis* have been characterized (Monnot *et al.* 2022). Most race-specific genes are of high resistance, uncomplicated inheritance, and can easily integrate into cultivated variety to attract breeders. When deploying qualitative resistors in cultivated variety, a different set of qualitative resistors must be used while monitoring the specific configuration of the targeted area, since only some of these resistance genes are effective when virulence concentrations in the microbial community are limited or absent. Breeders may use isolates with known virulence to evaluate the proportion of *R* genes in their breeding populations and chose resistances impacted by functional genes. Few *R* genes remained functional throughout time (Ray *et al.* 2022).

Growing breeding populations in as many sites as possible and host genotypes are easy ways to choose plant resistances. Using a range of isolates, the same genotypes should be checked for seedling susceptibility. Growing variations within the same experiment indicate which *R* genes are still functioning at each destination (Ijaz *et al.* 2022).

Quantitative Resistances

The defense mechanisms against plant diseases are intricate and multilayered. Traditionally, plant immunization has indeed been broadly classified into two groups: comprehensive resistance facilitated by resistance (*R*) proteins and imperfect resistance offered by QDR genes (Murugaiyan *et al.* 2022). In more recent days, the concept of 'microorganisms provoked immunity' has emerged in response to the role of microorganism's inducers and their host neurotransmitters (MTI). The zig-zag model of plant immunogenicity accounts for the latter two categories by combining generic pathogen inducers, their host neurotransmitters and *R*-protein-mediated resistance into a single product. In this case, plants' first line of defense is recognizing microbial elicitors such as flagellin and chitin. These microbial molecular patterns (MAMPs) are recognized by pattern-matching receptors (PRR) on the cell surface, which sets off a signaling cascade that often ends in inadequate defense responses and MTI (Iskandar *et al.* 2022).

Some infections are able to overcome this defense by secreting effector proteins that interfere with the host's metabolism, hence boosting the pathogen's infectiousness. Plants have evolved a specific kind of immunogenicity known as 'Effector-Triggered Immunity' (ETI) to detect and counteract the harmful effects of environmental stressors. Rigidity without parasite infection is a common side effect of ETI. Plant proteins that identify effectors are examples of *R* proteins, which are members of the family of proteins known as nucleotide-binding site leucine-rich receptors (NBS-LRRs). Instead of being diametrically opposed part of the immune system, MTI and ETI are now thought of as consistent output signals from the immune system resulting from differences in individual IPs and IPTRs. These models include measures of disease resistance (QDR) (Riseh *et al.* 2022).

Unlike MTI and ETI, QDR has received some previous recognition, although it is still widely misunderstood. QDR refers to host plant resistance that leads to a reduction in disease prevalence without leading to disease eradication. QDR is controlled by a network of genes that may function in combination with both environmental factors and other genes. However, recent studies have shown that pyramiding many QDR loci might result in substantial resistance. QDR has a statistical distribution for phenotypic resistivity values that is not consistent with Mendelian segregation ratios. The opposite is true for qualitative traits, where variation is almost always due to changes at a single locus and where the effects of different alleles at the locus are disproportionately large (Haider *et al.* 2023).

It is vital to remember that the genes regulating QDR are not the only things that might cause a population's reactance phenotype to shift. Recent studies have shown that QDR may arise through a variety of distinct mechanisms, including observable variation within the constituent parts of both MTI and ETI. This supports the idea that plant immunity occurs on a continuum from disease to resistance, with demonstrable variation in both pathogenic inducers and host responses, it also accords with the hypothesis behind the Invasion Model. The discovery that QDR is present in several pathosystems, such as *arabidopsis-R. solanacearum* and Arabidopsis-*Pseudomonas syringae*, lends credence to this theory (Harish *et al.* 2022). The term 'acceptance' refers to the host's ability to endure a stress pathogen load with little illness symptoms or strength and conditioning costs. Statistical resistance alleles give wide, long-lasting resistance. Pyramiding many resistance QTL may result in considerable disease resistance. A single resistance allele has a small influence on impedance. As indicated, maize has less resistance genes than other cereals. Maize possesses several dQTLs (dispersion QTLs) connected to resistance to necrotrophic pathogen-caused illnesses. By utilizing phenotypic or marker-assisted selection (MAS), resistance alleles from distinct dQTLs may be combined. Resistance can only be tested under pathogen-stressed settings that vary over time and location, unlike physical traits. Phenotypic selection is hampered (Parvin, 2022). For MAS to be effective, however, a thorough understanding of the genetic architecture regulating resistance to a specific illness is required. There are several examples of MAS in the literature that apply to QDR problems. A resilience allele of the real head smut dQTL and qHSR1 has been introduced using MAS into ten maize progenies with yield promise. While maintaining most of their other agronomic characteristics, all ten altered inbred lines and the hybrids they generated showed considerable increases in tolerance to head smut. Heterosis in hybrids is best used in a condition of genome-wide heterozygosity (Wu *et al.* 2022).

For the dominant resistance gene to take effect, it must be sufficient to introduce the resistance allele into only one of the hybrid's breeding lines (s). As a result, the hybrid may retain its enhanced resistance capabilities associated with its heterozygous state in the resistance-encoding genes. However, in the case of recessive resistance alleles, the gene must be present in both the parental lines for the offspring to show improved resistance performance (Narváez-Barragán *et al.* 2022). This need for homozygosity in a region may reduce heterosis if the inbred region transmitting the resistance allele is large. In this case, selecting the appropriate recombinants to reduce the size of the inbred area may be of paramount importance. It is feasible that parents might be bred with overlapping introgressions that contain the resistance gene, thus reducing the homozygous area. However, since maize hybrids are often improved, the converted resistant parental lines may become useless once the hybrid is no longer utilized economically (Skaliter *et al.* 2022). So, the dQTL allele/gene has to be introduced into a breeding population as soon as feasible. Note that recessiveness and dominance are not properties of a single gene, but rather of a pair of alleles sharing the same genome; hence, an allele that is recessive in one context may be dominant in another. Rough dwarf disease and southern rust are two diseases that may randomly strike maize fields in the Yellow plain of China. Maintaining resistance to these diseases is crucial, since an outbreak of any one of them in certain years might completely wipe out the harvest of maize (Whitfield *et al.* 2022).

RESISTANCE SOURCES

The most important factor to consider when beginning a breeding program is the accessibility of resistance sources. The optimal solution is when there is already sufficient genetic difference in varieties or known breeding populations that are adjusted to the targeted area. Resistance breeding has a longstanding experience in conventional pathosystems, such as the European wheat/powdery mildew problem (Whitfield *et al.* 2022). Then, resistance supplies can be crossed straight into improved varieties without first needing to be tested for other agronomic traits. When resistance supplies can be obtained from nearby nations within the same climatic region, such as resistances for wheat/stem rust from Austria or Hungary for German cultivars, it is correspondingly simple, though it takes a little longer. Phonological advancement and crop yield may differ in this case, but this can be conquered by recurrent selection that comes after (Sera *et al.* 2022). When resistance comes from old varieties or conventional landraces, such as those imported from China or Japan and used in Europe or the USA, far more work is required. Then, it is frequently lacking in adjustment to climatic parameters (such as winter hardiness, earliness)

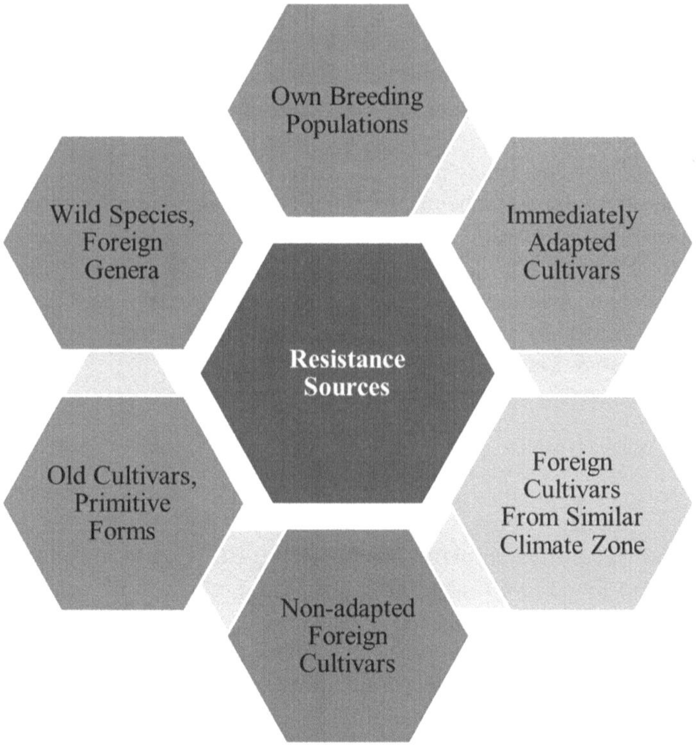

FIGURE 2.1 Sources of resistance.

and/or the production unit (such as plant height, yield level). When BaMMV first arrived in Germany in the 1970s, only an ancient Dalmatian landrace and Japanese cultivars were resistant (Keijzer *et al.* 2022). In the recent sunflower and potato cultivars, resistance from sexually crossed species, sometimes with embryo cultivation, is employed. Many rust-resistant strains in hexaploid wheat originate from wild progenitors (*Triticum dicoccoides*, *Triticum tauschii*), secondary (*Aegilops*, *Secale*) or tertiary gene pools (*Leymus villosum*) to establish late blight resistance in potatoes (Figure 2.1) (Gupta *et al.* 2022). Nonrelated host resistances does not survive as long as linked ones. Gene transformation technique provides the transfer of disease-resistant genes from non-hosts or other kingdoms (viruses and bacteria).

Biochemical Defense

The presence of dissolved silicon in plant cells may be related to a rise in resistance to fungi, according to the second hypothesis of silicon's biochemical enhancement of resistance. In this framework, increase in production of anti-fungal substances like phenolic various metabolic products (lignin), flavonoids, phytoalexins and pathogenesis-related proteins in plants, and initiation of some plant defense-related genes are the causes of the enhancement of resistance (Alayafi, 2022). These substances include increased activity of defense-related enzymes in leaves, such as polyphenol oxidase, peroxidase, phenylalanine ammonia-lyase and glucan. Many plants established stronger defense against additional pathogen attacks after spreading necrotizing pathogens—a phenomenon known as systemic obtained resistance (SOR). Due to silicon application on seedlings, two methods that increase the function of enzymes and anti-fungal substances may also cause SAR-like defense responses (Riaz *et al.* 2022). Additionally, other biochemical and physiological processes may play a role in the silicon-mediated disease resistance of plants. For instance, it has been observed that silicon probiotics can increase salicylic acid, jasmonic acid and ethylene levels in some host–pathogen interactions, such as the powdery mildew of arabidopsis induced by *Golovinomyces cichoracearum* and the rice-brown spot caused by *Cochliobolus miyabeanus* (Xiao & Liang, 2022).

Defense-Related Enzymes

Enzymes involved in defense are crucial for disease resistance. Numerous studies suggested that higher defensive enzyme activity was linked to lower disease frequency in the silicon-treated plants (Gulzar *et al.* 2022). After a fungal infection, silicon has been found to promote the concentration of defensive scheme enzymes in plant leaves. application of silicon in roots increased the activities of chitinase, peroxidases and polyphenol oxidases in cucumber plants affected by *Pythium* spp. *Mycosphaerella pinodes* were less common in pea seeds supplied with potassium silicate due to increased chitinase and -1,3-glucanase activity (Pazdiora *et al.* 2022). Wheat leaves treated with silicon had increased peroxidase activity, which lessened the magnitude of powdery mildew brought on by *B. graminisf.* Sp. *tritici*. Related to the interaction between *M. oryzae* and rice, higher concentrations of glucanase, peroxidase, polyphenol oxidase and phenylalanine ammonia-lyase are indicating increased resistance against the blast pathogen. According to Liang *et al.* (2005), silicon root implementation increased peroxidase, polyphenoloxidase and chitinase actions, which were effective in decreasing the magnitude of powdery mildew in cucumber (Liang & Maxworthy, 2005). Enhanced polyphenol oxidase activity lessened symptom severity in fruit handled with 1% silicon in relation to the interaction between *Cryptococcus laurentii* and sweet cherries. Increased peroxidase activity in sodium silicate-treated melon plants reduced the chance of pink rot brought on by *Trichothecium roseum* (Liang *et al.* 2022).

Recessive Resistances and Host Factors Required for Viral Infection

For the molecular pathways underlying recessive microbial resistance, alternative hypothesis can be put forth. The first implies that a blocker or a negative regulator of plant defense is involved. This situation has been proven for the *mlo* gene, which provides resistance to powdery mildew in barley (Silva & Fontes, 2022). Moreover, a gene homologous to the *A. thaliana CPR5* gene, known to be a defense mechanism controller, has recently been linked with RYMV2—a rice recessive resistance to the Rice Yellow Mottle Virus (RYMV). However, the widely accepted theory contends that a host factor specifically necessary for the achievement of the infection cycle was lost or mutated, leading to recessive resistance. This theory applies to viruses, because of the relatively low information content of viral genomes and the consequent requirement to enlist host cellular elements in order to complete the virus-induced cycle (Cai *et al.* 2022). Because of the complex interactions between the genes embedded by the host and viral genomes, plant–virus adaptability varies depending on each stage of the virus cycle. For infections between viruses and their hosts in plants, over 80% of plant virus resistances that have been reported are monogenically controlled. Recessive resistance genes are overexpressed in the interaction between plants and viruses when compared to other groups of pathogens (roughly 50% of the known resistance genes behave in a recessive manner). In sharp contrast to bacterial or fungal resistance, where the majority of genes are recessive, viruses have a relatively high proportion of these genes. Additionally, *potyviruses*, the largest and most economically damaging genus of plant viruses, are protected by the majority of recessive resistances found in crops. Therefore, it is not remarkable that research on plant–potyvirus interactions has produced the majority of the latest discoveries on recessive virus resistance (Andreakos *et al.* 2022).

Plant Resistances Mediated by the Translation Factor eIF4E

Potyviruses have such a single-stranded positive RNA genome that is covalently linked to a VPg (viral protein genome-linked) at its 5′ end and polyadenylate at its 3′ end. One of the viral predictors that influence plant–virus suitability has been identified as the VPg (Pankaj *et al.* 2022). The eukaryotic translation initiation factor 4E (eIF4E) and/or its isoform (eIF (iso) 4E) have been shown to interact with the VPg (or its NIa precursor) from several *potyviruses* in yeast two-hybrid and in in vitro binding assays. This interaction is critical for virus infectiousness in plants and is required for the formation of the *VPg–eIF*4E complex. These findings led researchers to search for a genetic connection between the eIF4E genes and recognized recessive resistance genes in crop species (Gao *et al.* 2022). A ‘candidate gene’ approach has shown over the past 10 years that many naturally occurring recessive resistant genes and alleles of the *eIF*4E or *eIF* (iso), 4E genes found in many species of plants. Further demonstrating the necessity of these cellular components for effective *potyvirus* infection, analysis of gene silencing (KO) and downregulation (silenced) plants impacted in their *eIF*4E and/or *eIF* (iso) 4E expression of genes showed that they were resistant to viral infection (Naseri *et al.* 2022). *Bymoviruses*, *Cucumoviruses*, *Ipomoviruses*, *Sobemoviruses*, *Carmoviruses* and *Waikiviruses* have all been included rapid expansion to other virus families, implying that *eIF*4E appears to contribute to a broad method of plant receptivity to viruses. The fact that *eIF*4E-mediated resistors encompass a wide variety of resistance phenotypes is an interesting feature of these resistors (González-Catrilelbún *et al.* 2022).

Using Molecular Approaches to Develop Disease Resistance in Plants

More than 30%–40% of global crop loss is caused by phytopathogens like viruses, bacteria and fungi, making them a major threat to agricultural productivity. To keep up with the increasing demand for food from a growing population, it is crucial to create disease-resistant crop varieties. Many methods have been used in the creation of disease-resistant crops (Javed *et al.* 2022). One of them is a technique for changing the genome called CRISPR-Cas9, which has fundamentally altered the approach that is taken to doing research on plants. The CRISPR genome editing (GE) approach may be used to develop plant diseases that are resistant to fungus, bacteria and other pathogens. In host plants, researchers have identified a number of genes that contribute to disease susceptibility, including recessive genes, genes that are negatively regulated and transcription factors. These genes and factors have the potential to perform some function in the plants defense mechanisms against infections. These genes were modified in many different plant species, including rice, wheat, cucumbers, cassava, cacao, grapes, citrus fruits, apples and bananas, among others, in order to enhance the plants' resistance to disease (Sagar *et al.* 2022).

Considering the rapid and ever-changing nature of viral infections, the management of viral diseases is very challenging. Many scientists have achieved significant advancement in the development of plant resistance via the creation of viral and non-viral proteins, host resistant strains and gene silencing approaches using RNA interference (RNA interference) (Uyttebroek *et al.* 2022). Because of the benefits of CRISPR, there has been a significant increase in the production of plants that are susceptible to DNA and RNA viruses. Directly addressing the viral genome and focusing on plant recessive 'S' genes, both of which make it easy for viruses to replicate in plants, are the two primary approaches that are being investigated in order to engineer virus resistance. The viral genomes were modified via the use of the CRISPR-Cas9 technology in order to increase plant tolerance to viral infections (Apinda *et al.* 2022) (Table 2.1).

There are variants of the *Geminiviridae* virus family that pose a risk to the harvest of almost all crops grown world widely. The rolling circle technique of amplification

TABLE 2.1
Difference between Qualitative and Quantitative Resistances

Sr.	Features	Quantitative Resistance	Qualitative Resistance
1	Phenotypic expression	Horizontal resistance	Vertical resistance
2	Pathotype/specificity	Race-nonspecific	Race-specific
3	Response to the pathogen	Show resistance	Hypersensitive
4	Durability	High	Relatively low
5	Expression stage	As plant mature expression increase (adult plant resistance)	Seedling to maturity (all stages resistance)
6	Nature of gene	Polygenic	Oligogenic
7	Disease symptoms	Vary from disease to disease	No disease
8	Breeding strategy	Recurrent and multistage selection	Crossing and back crossing
9	Efficiency	Variable	Highly efficient
10	Assessment and selection	Depending on disease severity and relatively difficult	Depending on infection type or relatively easy
11	Resistance genes	ZmWAK, Htn1, Yr36, RFO1/ WAKL22	Yr6, Yr17, Yr32, Yr8, Yr10, Yr24, Yr5, sp, Yr15

(RCA) makes it possible for viruses with DNA A and B genomes that are single-stranded to multiply into DNA that is double-stranded within the nuclei of host plants (Radouane *et al.* 2022). In the first experiment, *N. benthamiana* and *A. thaliana* were altered to have tolerance to the beet severe curly top virus (BSCTV) and the yellow dwarf virus (BeYDV) by the use of CRISPR-mediated viral resistance (Dudley *et al.* 2022). When the gRNAs targeting yellow dwarf virus' (YDV) genes that code for reproduction and cell mobility were highly expressed, plants exhibited an increased level of resistance to the yellow dwarf virus (YDV). CRISPR-Cas9 dual gRNAs targeting the C1 and IR sections of Cotton Leaf Curl Multan Virus (CLCuMuV) were able to exhibit 100% resistant to leaf curl disease and prevent viral infections in mutant tobacco plants (Talakayala *et al.* 2022). In *N. benthamiana* targeting many viral genome regions, such as intergenic (TAATATTAC), coding and noncoding regions, may be an effective method for controlling CLCuV, TYLCV, beet curly top virus (BCTV) and *Merremia mosaic virus* (MeMV) (Khan *et al.* 2019; Gebre *et al.* 2022).

Fungal infections are the most common cause of disease in a wide range of food crops, including wheat and corn (Shukla *et al.* 2022). Because the pathogen-effector proteins that are secreted by them have a high degree of polymorphism, no existing approach can establish long-term resistance. In the current age of crop production, the innovative CRISPR-Cas9 technology for controlling agricultural diseases is indispensable. Editing of '*S*' genes using CRISPR-Cas9 led to a diverse array of protection toward diseases caused by fungi and bacteria (Shukla *et al.* 2022). The 'S' genes assist the pathogens by facilitating infection and helping in the formation of a beneficial association. It is feasible that changes in the alleles of the '*S*' gene might create a greater degree of resilience than the resistance given by the *R* gene (Wang *et al.* 2021). Powdery mildew may infect both monocot and dicot plants, although dicots are more likely to be affected. CRISPR-/Cas9 was utilized to edit one of the key genes, *MLO*; as a consequence, wheat varieties are more resistant to powdery mildew infection (Mishra *et al.* 2022).

The continued interference of bacterial infections in agricultural production and output poses a risk to the safety of food supplies across the world. Numerous '*S*' genes in plants are engaged in the interaction between both the host and pathogenic bacteria; hence, these genes are novel targets for gene editing that may confer resistance to bacterial diseases. Recently through TALENs the DNA of *SWEET* genes have been modified (Andersson *et al.* 2022). *Xanthomonas oryzae* PV is responsible for a number of devastating diseases, one of which is bacterial blight of rice. In order to invade and pathogenic its plant host, Xoo secretes TALE proteins into host cell membranes and induces the expression of sucrose efflux transporters (SWEET genes). By disrupting the communication between bacterial TALE effector proteins and plant *SWE*ET genes, plant pathogen development, and pathogenicity may be suppressed. Furthermore, researchers used TALENs to target the effector binding sites (EBSs) in the promoters of *SWE*ET genes, therefore establishing resistance against bacterial blight (Andersson *et al.* 2022). The scientific communities of different countries should establish ethical and regulatory criteria and translate genome-edited crops for the betterment of human health (Ji *et al.* 2022).

FUTURE PERSPECTIVES

Disease resistance in plants is a crucial matter needs serious consideration. In nature, certain plant varieties and cultivars have innate immunity to common plant pathogens. Some have a waxy covering on their surface that resist harmful bacteria. Some plants can increase their resistance to disease by reacting to environmental variables that boost the plant's immune system. Unfortunately, plant breeding is a lengthy process, and both introducing resistant cultivars and transferring genes to non-resistant cultivars is difficult. It's also important to think about whether the imported resistant cultivars will thrive in the region. Advantageous bacteria and non-pathogenic variants have been used to successfully remove several insect pests in greenhouses and labs. Furthermore, many drugs may not function as expected in the field because of the complex interactions between the surroundings, microorganisms, plants and biological variables (e.g., PGPR). As a result, ecofriendly abiotic activator chemicals that may induce plant resistance in stressful environments should be emphasized in plant disease resistance research. CRISPR-mediated base editing and prime editing are also promising new methods for the development of resistance to disease. It helps researchers create several new varieties with enhanced traits that are resistant to diseases. Polyploid plants, in which the identification and differentiation of mutants is crucial, have been modified genetically. By using this technique, it is possible to verify the effectiveness of mutated lines in record time. By analyzing the capability of certain crop species, boosting their resilience to vulnerable pests and diseases, sustaining productivity, enduring abiotic stress and enhancing nutritional effectiveness, we may find new opportunities for the farming of the future using CRISPR techniques. Potentially, this method might pave the way for the next green revolution by facilitating the creation of next-generation breeding techniques for sustainable agricultural production.

REFERENCES

Aboulila, A. A., & Pathology, M. P. (2022). Efficiency of plant growth regulators as inducers for improve systemic acquired resistance against stripe rust disease caused by Puccinia striiformis f. sp. tritici in wheat through up-regulation of PR-1 and PR-4 genes expression. *Physiological and Molecular Plant Pathology*, *121*, 101882.

Aitouguinane, M., Alaoui-Talibi, Z. E., Rchid, H., Fendri, I., Abdelkafi, S., El-Hadj, M. D. Traïkia, M. (2022). Polysaccharides from Moroccan green and brown seaweed and their derivatives stimulate natural defenses in olive tree leaves. *Applied Sciences, 12*(17), 8842.

Alayafi, A. H. (2022). *Effect of Irrigation Water Salinity and Silicon Application on Yield of Maize (Zea mays L.) and Water Use Efficiency under Trickle Irrigation System* (Doctoral dissertation, King Abdulaziz University). *Agronomy*, *11*(11), 2299.

Andersson, N., Saba, K. H., Magnusson, L., Nilsson, J., Karlsson, J., Nord, K. HCancer. (2022). Inactivation of RB1, CDKN2A and TP53 have distinct effects on genomic stability at side-by-side comparison in karyotypically normal cells. *Genes, Chromosomes and Cancer*, 62(2), 93–100.

Andreakos, E., Abel, L., Vinh, D. C., Kaja, E., Drolet, B. A., Zhang, Q., Haerynck, F. (2022). A global effort to dissect the human genetic basis of resistance to SARS-CoV-2 infection. *Nature Immunology*, 159–164.

Apinda, N., Yao, Y., Zhang, Y., Reddy, V. R., Chang, P., Nair, V., & Sthitmatee, N. (2022). CRISPR/Cas9 editing of duck enteritis virus genome for the construction of a recombinant vaccine vector expressing ompH gene of pasteurella multocida in two novel insertion sites. *Vaccines*, 10(5), 686.

Bozbuga, R., Arpaci, B. B., Uluisik, S., Guler, P. G., Yildiz, H. N., & Ates, S. Y. (2022). Genetic modification of plant hormones induced by parasitic nematodes, virus, viroid, bacteria, and phytoplasma in plant growing. *Plant Hormones: Recent Advances, New Perspectives and Applications,* 167.

Cai, L., Liu, J., Wang, S., Gong, Z., Yang, S., Xu, F., Yang, J. (2022). The coiled-coil protein gene WPRb confers recessive resistance to Cucumber green mottle mosaic virus. *Plant Physiology, 191*(1), 369–381.

Chiu, C. H., Roszak, P., Orvošová, M., & Paszkowski, U. (2022). Arbuscular mycorrhizal fungi induce lateral root development in angiosperms via a conserved set of MAMP receptors. *Current Biology*, *32*(20), 4428–4437.

Chow, E., Wang, K., Lim, C. K., Tsoi, S. T., Fan, B., Poon, E., Mari, A. (2022). Dorzagliatin, a dual-acting glucokinase activator, increases insulin secretion and glucose sensitivity in glucokinase-maturity-onset diabetes of the young (GCK-MODY) and recent-onset type 2 diabetes. *Diabetes*, *72*(2), 299–308.

D'Andrea, R., & Dullerud, G. E. (2003). Distributed control design for spatially interconnected systems. *IEEE Transactions on Automatic Control*, *48*(9), 1478–1495.

Deng, C., Leonard, A., Cahill, J., Lv, M., Li, Y., Thatcher, S., Li, Z. (2022). The RppC-AvrRppC NLR-effector interaction mediates the resistance to southern corn rust in maize. *Molecular Plant*, *15*(5), 904–912.

Dudley, Q. M., Jo, S., Guerrero, D. A. S., Chhetry, M., Smedley, M. A., Harwood, W. A., Patron, N. (2022). Reconstitution of monoterpene indole alkaloid biosynthesis in genome engineered Nicotiana benthamiana. *Communications Biology*, *5*(1), 949.

Ebrahimi-Zarandi, M., & Skorik, Y. (2022). Alginate-induced disease resistance in plants. *Polymers*, *14*(4), 661.

Ezati, P., Priyadarshi, R., Rhim, & Technologies. (2022). Prospects of sustainable and renewable source-based carbon quantum dots for food packaging applications. *Sustainable Materials and Technologies*, e00494.

Fosha, C. E., Elgerd, O. I. & Systems. (1970). The megawatt-frequency control problem: A new approach via optimal control theory. *IEEE Transactions on Power Apparatus and Systems*, (4), 563–577.

Gao, H., Ma, K., Ji, G., Pan, L., Wang, Z., Cui, M., & Zhou, Q. (2022). Protein glycosylation changes during systemic acquired resistance in Arabidopsis thaliana. *International Journal of Biological Macromolecules*, *212*, 381–392.

Gao, J., Tang, Y.-D., Hu, W., & Zheng, C. (2022). When poly (A) binding proteins meet viral infections, including SARS-CoV-2. *Journal of Virology*, *96*(7), e00136–22.

Gebre, M. S., Rauch, S., Roth, N., Yu, J., Chandrashekar, A., Mercado, N. B., Martinot, A. (2022). Optimization of non-coding regions for a non-modified mRNA COVID-19 vaccine. *Nature*, 601(7893):410–414.

Golea, D. G., & Hideg, C. (2022). *Bioterrorism* (Vol. 1): Introduction to bioterrorism: Origin, manifestation, causes, economic and substratum, Lucian Ştefan COZMA (Editor). Golea Daniela Georgiana, 5–519.

González-Catrilelbún, S., Cartagena, J., Vargas, D., Breguel-Serrano, P., Sandino, A. M., & Rivas-Aravena, A. (2022). The RNA-dependent RNA polymerase of the infectious pancreatic necrosis virus is linked to viral mRNA acting as a cap substitute. *Journal of General Virology*, *103*(3), 001729.

Gulzar, N., Lone, R., Kamili, A. N., Shah, M. A., Alaklabi, A., & Pathology, M. P. (2022). Expression of defense genes and free radical scavenging on pre supplementation of silicon to Alternaria solani inoculated Lycopersicon esculentum plants. *Physiological and Molecular Plant Pathology*, *123*, 101929.

Gupta, D. S., Kumar, J., Parihar, A. K., Chandra, A., Sujayanand, G., & Gupta, S. (2022). Urdbean breeding. In *Fundamentals of Field Crop Breeding*. Springer. pp. 1151–1179.

Haider, A., Ikram, M., & Rafiq, A. (2023). Introduction to veterinary bacteriology. In *Green Nanomaterials as Potential Antimicrobials*. Springer. pp. 87–108.

Harish, V., Tewari, D., Gaur, M., Yadav, A. B., Swaroop, S., Bechelany, M., & Barhoum, A. J. (2022). Review on nanoparticles and nanostructured materials: Bioimaging, biosensing, drug delivery, tissue engineering, antimicrobial, and agro-food applications. *Nanomaterials, 12*(3), 457.

He, Z., Webster, S., & He, S. Y. (2022). Growth–defense trade-offs in plants. *Current Biology*, 32(12): 634–639.

Ijaz, U., Pervaiz, T., Ahmed, T., Seemab, R., Shahid, M., Noman, M., Nadeem, M. & Azeem, F. (2020). AJAB. *Asian Journal of Agriculture and Biology*, *8*(2), 207–222.

Iskandar, K., Hawser, S., Hays, J., Mohsen, Y., Adukkadukkam, S., Awuah, W., & Jose, R. (2022). Progress in alternative strategies to combat antimicrobial resistance: Focus on antibiotics. *Antibiotics, 11* 200. In: s Note: MDPI stays neutral with regard to jurisdictional claims in. *Antibiotics, 11*(2), 200.

Javed, T., Indu, I., Singhal, R. K., Shabbir, R., Shah, A. N., Kumar, P., Saha, D. (2022). Recent advances in agronomic and physio-molecular approaches for improving nitrogen use efficiency in crop plants. *Frontiers in Plant Science, 13.*

Jeandet, P., Lahlali, R., & Barka, E. A. (2022). The hidden world within plants. In (Vol. 10, pp. 1951): MDPI. *Microorganisms, 10*(10), 1951.

Ji, Z., Guo, W., Chen, X., Wang, C., & Zhao, K. (2022). Plant executor genes. *International Journal of Molecular Sciences, 23*(3), 1524.

Jiang, Y., Ji, X., Zhang, Y., Pan, X., Yang, Y., Li, Y., Products. (2022). Citral induces plant systemic acquired resistance against tobacco mosaic virus and plant fungal diseases. *Industrial Crops and Products, 183*, 114948.

Keijzer, P., van Bueren, E., Engelen, C., & Hutten, R. (2022). Breeding late blight resistant potatoes for organic farming—a collaborative model of participatory plant breeding: The Bioimpuls Project. *Potato Research*, 1–29.

Kevan, P., Chaloner, W., & Savile, D. (1975). Interrelationships or early terrestrial arthropods and plants. *Palaeontology, 18*(2), 391–417.

Khan, A., Wadood, S. F., Chen, M., Wang, Y., Xie, Z.-P., & Staehelin, C. (2022). Effector-triggered inhibition of nodulation: a rhizobial effector protease targets soybean kinase GmPBS1-1. *Plant Physiology, 189*(4), 2382–2395.

Lazazzara, V., Avesani, S., Robatscher, P., Oberhuber, M., Pertot, I., Schuhmacher, R., & Perazzolli, M. (2022). Biogenic volatile organic compounds in the grapevine response to pathogens, beneficial microorganisms, resistance inducers, and abiotic factors. *Journal of Experimental Botany, 73*(2), 529–554.

Liang, D., Huang, J., Zhang, J., Shi, S., Zhu, N., Chen, & Engineering. (2022). Three-dimensional simulations of scour around pipelines of finite lengths. *Journal of Marine Science and Engineering, 10*(1), 106.

Liang, H., & Maxworthy, T.. (2005). An experimental investigation of swirling jets. *Journal of Fluid Mechanics, 525*, 115–159.

Mendoza-de Gives, P. (2022). Soil-borne nematodes: Impact in agriculture and livestock and sustainable strategies of prevention and control with special reference to the use of nematode natural enemies. *Pathogens, 11*(6), 640.

Min, C. W., Jang, J. W., Lee, G. H., Gupta, R., Yoon, J., Park, H. J., Cho, L., (2022). TMT-based quantitative membrane proteomics identified PRRs potentially involved in the perception of MSP1 in rice leaves. *Journal of Proteomics, 267*, 104687.

Mishra, J., Mishra, I., & Arora, N. (2022). 2, 4-Diacetylphloroglucinol producing Pseudomonas fluorescens JM-1 for management of ear rot disease caused by Fusarium moniliforme in Zea mays L. *3 Biotech, 12*(6), 138.

Monnot, S., Cantet, M., Mary-Huard, T., Moreau, L., Lowdon, R., Van Haesendonck, M., Boissot, N. (2022). Unravelling cucumber resistance to several viruses via genome-wide association studies highlighted resistance hotspots and new QTLs. *Horticulture Research. 9.*

Murugaiyan, J., Kumar, P. A., Rao, G. S., Iskandar, K., Hawser, S., Hays, J. P., Jose, R. A. (2022). Progress in alternative strategies to combat antimicrobial resistance: Focus on antibiotics. *Antibiotics*, *11*(2), 200.

Narváez-Barragán, D. A., Tovar-Herrera, O. E., Guevara-García, A., Serrano, M., & Martinez-Anaya, C. (2022). Mechanisms of plant cell wall surveillance in response to pathogens, cell wall-derived ligands, and the effect of expansins to infection resistance or susceptibility. *Frontiers in Plant Science*, *13*.

Naseri, A., Moradi, Z., Mehrvar, M., Zakiaghl, M., & Technology. (2022). Genomic characterization and phylogenetic analysis of two potyviruses infecting iris in Iran. *Journal of Agricultural Science and Technology*, *24*(5), 1233–1249.

Nguyen, N. H., Trotel-Aziz, P., Villaume, S., Rabenoelina, F., Clément, C., Baillieul, F., & Aziz, A. (2022). Priming of camalexin accumulation in induced systemic resistance by beneficial bacteria against Botrytis cinerea and Pseudomonas syringae PV. Tomato DC3000. *Journal of Experimental Botany*, *73*(11), 3743–3757.

Oladosu, Y., Rafii, M. Y., Arolu, F., Chukwu, S. C., Salisu, M. A., Fagbohun, I. K., Haliru, B. (2022). Superabsorbent polymer hydrogels for sustainable agriculture: A review. *Horticulturae*, *8*(7), 605.

Osuna-Caballero, S., Rispail, N., Barilli, E., & Rubiales, D. (2022). Identification and characterization of novel sources of resistance to rust caused by uromyces pisi in Pisum spp. *Plants*, *11*(17), 2268.

Pankaj, Y. K., Kaur, R., Kumar, R., Kumar, V., Razzaq, A., & Wani, S. H. (2022). Mitigation of soybean mosaic virus using an efficient molecular approach. In Soybean Improvement. *Springer.* pp. 183–201.

Parvin, P. (2022). Hidden in plain sight: How lobby organisations undermine democracy. In Wealth and Power. *Routledge*. pp. 229–251.

Pazdiora, P. C., Crizel, R. L., da Rosa Dorneles, K., da Silva, S. M., Morello, T. N., Chaves, F. C., & Dallagnol, (2022). Silicon amendment improves wheat defense against Fusarium graminearum and complements the control by fungicide of Fusarium head blight. *Plant Pathology*, 53–66.

Pereira, F. B., & González-Solís, D. (2022). Review of the parasitic nematodes of marine fishes from off the American continent. *Springer*. pp. 1–60.

Radouane, N., Lahlali, R., Darif, M., Ezrari, S., Benjelloun, M., Belabess, Z., Tahiri, A. (2022). Detection and molecular characterization of chickpea Chlorotic dwarf virus and tomato leaf curl new Delhi virus in morocco. *Horticulturae*, *8*(10), 927.

Ramírez-Zavaleta, C. Y., García-Barrera, L. J., Rodríguez-Verástegui, L. L., Arrieta-Flores, D., & Gregorio-Jorge, J. (2022). An overview of PRR-and NLR-mediated immunities: Conserved signaling components across the plant kingdom that communicate both pathways. *International Journal of Molecular Sciences*, *23*(21), 12974.

Ray, A., Gowda, G. B., Patil, N. K. B., Pandi, G. P., Adak, T., Pradhan, H., Rath, P. C. (2022). Advances in insect resistance breeding against brown planthopper and gall midge in rice. In *Genetic Methods and Tools for Managing Crop Pests*. Springer. pp. 641–653.

Riaz, S., Hussain, I., Parveen, A., Arshraf, M. A., Rasheed, R., Zulfiqar, S., Rehman, S. (2022). Silicon and nano-silicon in plant nutrition and crop quality. In *Silicon and Nano-silicon in Environmental Stress Management and Crop Quality Improvement. Elsevier*. pp. 277–295.

Riseh, R. S., Hassanisaadi, M., Vatankhah, M., Soroush, F., & Varma, R. (2022). Nano/micro-encapsulation of plant biocontrol agents by chitosan, alginate, and other important biopolymers as a novel strategy for alleviating plant biotic stresses. *International Journal of Biological Macromolecules*, 1589–1604.

Rothman, A. A., & Wagner, (2003). Chronic illness management: What is the role of primary care? *Annals of Internal Medicine*, *138*(3), 256–261.

Riseh, R. S., Gholizadeh Vazvani, M., Ebrahimi-Zarandi, M., & Skorik, Y. (2022). Alginate-induced disease resistance in plants. *Polymers*, *14*(4), 661.

Sagar, A., Rai, S., Ilyas, N., Sayyed, R., Al-Turki, A. I., El Enshasy, H. A., & Simarmata, T. (2022). Halotolerant rhizobacteria for salinity-stress mitigation: Diversity, mechanisms and molecular approaches. *Polymers*, *14*(4), 661.

Sandell, N., Varaiya, P., & Athans, M., US Dept. of Commerce. (1975). A survey of decentralized control methods for large scale systems. *Systems Engineering for Power, US Dept. of Commerce*, 334–335.

Sanguankiattichai, N., Buscaill, P., & Preston, G. (2022). How bacteria overcome flagellin pattern recognition in plants. *Current Opinion in Plant Biology*, *67*, 102224.

Sera, G. H., de Carvalho, C.., de Rezende Abrahão, J. C., Pozza, E. A., Matiello, J. B., de Almeida, S. R., dos Santos Botelho, (2022). Coffee leaf rust in Brazil: Historical events, current situation, and control measures. *Agronomy*, 12(2), 496.

Sett, S., Prasad, A., & Prasad, M. (2022). Resistance genes on the verge of plant–virus interaction. *Trends in Plant Science*, 1242–1252.

Sharma, H., Haq, M. A., Koshariya, A. K., Kumar, A., Rout, S., & Kaliyaperumal, K. (2022a). Research article "Pseudomonas fluorescens" as an antagonist to control okra root rotting fungi disease in plants *Journal of Food Quality*, 5608543.

Sharma, H., Haq, M. A., Koshariya, A. K., Kumar, A., Rout, S., & Kaliyaperumal, K. (2022b). "Pseudomonas fluorescens" as an antagonist to control okra root rotting fungi disease in plants. *Journal of Food Quality*, 5608543.

Shayeghi, H., Shayanfar, H., Jalili, A., & management. (2009). Load frequency control strategies: A state-of-the-art survey for the researcher. *Energy Conversion and Management*, *50*(2), 344–353.

Shokrollahi, N., Ho, C.-L., Mohd Zainudin, N., Wahab, A., Bin, M. A., & Wong, M.-Y. (2022). Plant defense inducers and antioxidant metabolites produced during oil palm-ganoderma boninense interaction in vitro. *Springer*, 1–13.

Shukla, S., Upadhyay, D., Mishra, A., Jindal, T., & Shukla, K. (2022). Challenges faced by farmers in crops production due to fungal pathogens and their effect on indian economy. In *Fungal Diversity, Ecology and Control Management.* Springer. pp. 495–505.

Silva, F. D., & Fontes, E. P (2022). Clustered regularly interspaced short palindromic repeats-associated protein system for resistance against plant viruses: Applications and perspectives. *Transgenic Research, 31*(2), 167–199.

Skaliter, O., Livneh, Y., Agron, S., Shafir, S., & Vainstein, A. (2022). A whiff of the future: functions of phenylalanine-derived aroma compounds and advances in their industrial production. *Plant Biotechnology Journal*, *20*(9), 1651–1669.

Smith, S. (2007). An overview of ecological and habitat aspects in the genus Fusarium with special emphasis on the soil-borne pathogenic forms. *Plant Pathology Bulletin, 16*, 97–120.

Soto, W. (2022). Emerging Research Topics in the Vibrionaceae and the Squid–Vibrio Symbiosis. *Microorganisms, 10*(10), 1946.

Sun, Z., Cheng, Y., Qi, F., Zhang, M., Tian, M., Wang, J., Zhang, X. (2022). Resistance of peanut to web blotch caused by Phoma arachidicola is related to papillae formation and the hypersensitive response. *Plant Pathology*, *71*(9), 1921–1931.

Swain, B. B., Mohapatra, P. K., Naik, S. K., & Mukherjee, A. (2022). Biopriming for induction of disease resistance against pathogens in rice. *Planta*, *255*(6), 113.

Talakayala, A., Ankanagari, S., & Garladinne, M. (2022). Harnessing CRISPR/Cas Tools for Installing Virus Resistance in Cereals: An Overview. *Next-Generation Plant Breeding Approaches for Stress Resilience in Cereal Crops*, 433–452.

Thankappan, S., Narayanasamy, S., Sridharan, A., Binodh, A., Kumari, A. N., Parasuraman, P., Pathology, M. P. (2022). Rhizospheric volatilome in modulating induced systemic resistance against biotic stress: A new paradigm for future food security. *Physiological and Molecular Plant Pathology*, 101852.

Uyttebroek, S., Chen, B., Onsea, J., Ruythooren, F., Debaveye, Y., Devolder, D., Lavigne, R. (2022). Safety and efficacy of phage therapy in difficult-to-treat infections: A systematic review. *The Lancet Infectious Diseases*, (8), e208–e220.

Venkat, A. N., Hiskens, I. A., Rawlings, J. B., & Wright, S. (2008). Distributed MPC strategies with application to power system automatic generation control. *IEEE Transactions on Control Systems Technology*, *16*(6), 1192–1206.

Wang, Y., Ji, L., Li, Q., Wu, Y., Li, C., Kong, L., & Wang, X. (2021). Effects of long-term bare fallow during the winter-wheat growth season on the soil chemical properties, fungal community composition, and the occurrence of maize fungal diseases in North China. *Plant Disease*, *105*(9), 2575–2584.

Wei, W., Xu, L., Peng, H., Zhu, W., Tanaka, K., Cheng, J., Chen, W. (2022). A fungal extracellular effector inactivates plant polygalacturonase-inhibiting protein. *13*(1), 1–15. *Nature Communications*, *13*(1), 2213.

Whitfield, C., Szymanski, C. M., Lewis, A. L., & Aebi, M. (2022). Eubacteria. In *Essentials of Glycobiology* [Internet]. 4th edition. Cold Spring Harbor (NY): Cold Spring Harbor Laboratory Press, 21, pp. 1–26.

Wu, Y.-Z., Zhang, Q., Wei, X.-H., Jiang, C.-X., Li, X.-K., Shang, H.-C., & Lin, S. (2022). Multiple anti-inflammatory mechanisms of Zedoary Turmeric Oil Injection against lipopolysaccharides-induced acute lung injury in rats elucidated by network pharmacology combined with transcriptomics. *Phytomedicine*, *106*, 154418.

Wulff, B. B., & Krattinger, S. (2022). The long road to engineering durable disease resistance in wheat. *Current Opinion in Biotechnology*, *73*, 270–275.

Xiao, Z., & Liang, Y (2022). Silicon prevents aluminum from entering root tip by promoting formation of root border cells in rice. *Plant Physiology and Biochemistry*, *175*, 12–22.

Yadav, N., Mudgal, D., Anand, R., Jindal, S., & Mishra, V. (2022). Recent development in nanoencapsulation and delivery of natural bioactives through chitosan scaffolds for various biological applications. *International Journal of Biological Macromolecules*, 1(11), 537–572.

Yu, Y., Gui, Y., Li, Z., Jiang, C., Guo, J., & Niu, D. (2022). Induced systemic resistance for improving plant immunity by beneficial microbes. *Plants*, *11*(3), 386.

Zhao, Y., Zhu, X., Chen, X., & Zhou, J. (2022). From plant immunity to crop disease resistance. *Journal of Genetics and Genomics*, *49*(8), 693–703.

3 CRISPR-Cas
An Arsenal against Plant Diseases

INTRODUCTION

By 2050, it is predicted that the worldwide need of food production would have doubled (Tomlinson *et al.* 2013). Since climate change is expected to affect food production by increasing dry spells and average daily temperatures in certain agronomically sensitive locations, this uncomfortable anticipation is becoming more relevant. All kind of plant life might be affected. Parasites, whether they are viruses, bacteria, fungi, parasitic plants, nematodes, have caused devastating losses due to plant diseases. Each year, plant irritations and diseases cause 20%–40% of crop losses (Panhwar *et al.* 2022). Microorganisms that infect plants have severely damaged the agriculture sector across the world. The virus infects both susceptible and tolerant plants in a similar and methodical fashion, but the tolerant plants are better able to combat the infection and slow its progress. Agriculturalists and plant pathologists face tremendous challenges. Conventional agricultural disease control still relies heavily on chemical pesticides (Teske & Nagrath, 2022).

However, there are many cases when these pesticides cause harm to humans and ecosystems. Pesticides may have unintended consequences on non-target organisms and pathogens, upsetting delicate ecological equilibrium. It is possible that pesticide resistance may emerge alongside hazardous pests, making it necessary to create new pesticides or resort to other methods of control (Wang *et al.* 2022). Therefore, it is essential, especially in developing countries, to reduce the dependency on chemical pesticides in food production in order to minimize negative environmental repercussions. Pesticides, natural predators, and physical barriers are the mainstays of conventional approaches to preventing the spread of infectious diseases caused by viruses. Disease control strategies include culturing techniques as well as early planting, weed reduction, crop-free intervals, virus-free propagation material, and the dumping of infected crops (Wudil *et al.* 2022).

Historically, it has been difficult to develop a reliable, long-term strategy for controlling viral disease outbreaks because of the complex epidemiological parameters involved in such outbreaks. These include, the dynamics of migrating vectors, the rapidity with which viruses evolve, and the extent to which the viruses can infect a wide variety of hosts. Breeding crops to be resistant to diseases is a sustainable farming practice that is also kind to the environment. In spite of its usefulness over the last several decades, traditional resistance breeding has had a few disadvantages (Copeland *et al.* 2022).

DOI: 10.1201/b22901-3

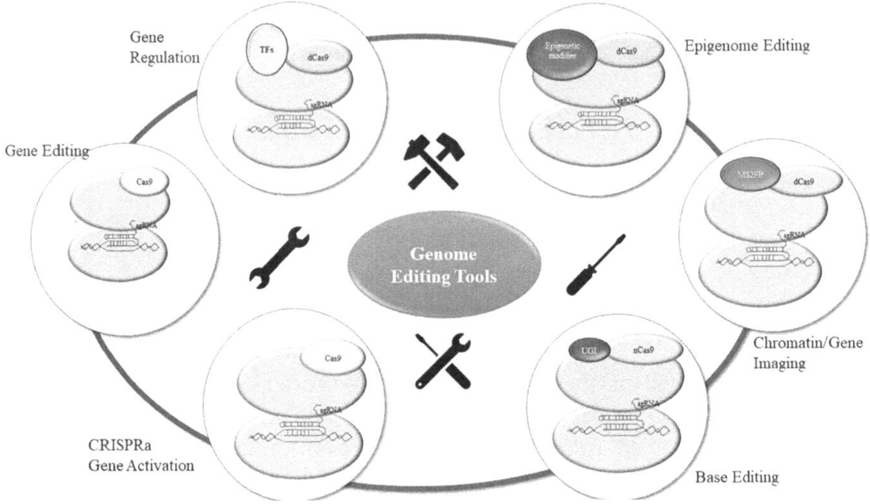

FIGURE 3.1 CRISPR toolbox.

CRISPR ROLE IN PLANT GENETIC IMPROVEMENTS

CRISPR technology has been used for various purposes in plants ranging from modifications at DNA, RNA, and epigenetic levels. CRISPR toolbox has various tools that may be employed to get desirable modifications in the plant genome (Figure 3.1).

Gene Knockout Based on CRISPR Technology

In efforts to facilitate fine-tuned modifications to the structure of the genome, researchers have turned to CRISPR-Cas9 technology. It may be used to learn more about the function of a certain gene or set of genes, change the physical appearance of the plant, increase its yield or make it more resistant to disease. At first, it was shown that the two *BnaMAX1* genes in rapeseed, which control plant height and lateral bud development, perform redundant tasks were targeted (Lin *et al.* 2022). Utilizing CRISPR-Cas9 technology, two short guide RNAs (sgRNAs) were designed to make edits to the *BnaMAX1* genes. Enhanced branch development and semidwarf stature were the results of the presence of all four *BnaMax1* alleles in rapeseed cultivars. When compared to their wild-type counterparts, these cultivars outperformed the others. Mutations at the *BnaMAX1* target loci in T0 plants were found to have a high success rate (56.30%–67.38%), including homozygous, heterozygous, chimeric, and biallelic mutations (Mianné *et al.* 2022).

Further research revealed that the mutations were transmitted to both T1 and T2 descendants. The rapeseed ideotype might be generated using these mutant variants. CRISPR was used to create a knockout mutation in the gene for granule-bound starch synthase in potato (GBSS) (Bazaz & Dehghani, 2022). This variant has an increased ability to produce amylopectin starch, making it more attractive to the food industry. Saika et al. (2019) used CRISPR technology to modify the Tos17 retrotransposon gene in rice. To eliminate the inter-LTR space in vitro, targeted mutagenesis of the LTR region was performed in calli cells. Success was also shown with plants regenerated from modified calli; these plants were characterized as homozygous for a *Tos17* knockout (Sukegawa *et al.* 2021).

Gene Knockin Based on CRISPR Technology

With homology-directed repair (HDR), exact gene replacement or knock in is possible, making it viable by a double-strand break (DSB) recovery approach. Although research on HDR is limited, researchers effectively inserted the AvrII sites of the *NbPDS* gene by HDR-mediated gene substitution by co-expressing Cas9 and gRNA in tobacco protoplasts (Saika *et al.* 2019).

Fertility Boosting and Nutritional Supplements

Metabolic processes, stress tolerance, nutritional production, and herbicide resistance are just a few of the areas where CRISPR technology has been applied to great effect. It was discovered that the dominant male sterile gene *Ms2* was located on the 4DS chromosome in a Chinese wheat cultivar. The offspring of a hybrid between normal fertile wheat and male sterile wheat types will be a mixture of 50% fertile and 50% male sterile (Eissa *et al.* 2020). The *Ms2* gene was modified by the CRISPR-Cas9 system regulated by agrobacterium, and male fertility in wheat was reintroduced. To increase the effectiveness of gene editing, two sgRNAs targeting exons IV and VII of the *Ms2* gene were developed. Infection of hybrid grain embryos with *Agrobacterium tumefaciens* carrying the sgRNA cloned to an expression vector was conducted. Ninety percent of attempts to alter *Ms2* were effective, resulting in restored fertility in dwarf male-sterile wheat (DMSW) (Mendoza *et al.* 2022).

Regulation of Transcription and Translation

CRISPR allows for the regulation of promoters, enhancers, and factors involved in transcription and translation. Changes in the host system's protein abundance are possible (Xia *et al.* 2022). By silencing the eukaryotic translation initiation factor gene *elF4E*, CRISPR was able to create disease-resistant plants that were resistant to cucumber vein yellowing virus, potyviruses, zucchini yellow mosaic virus, and papaya ring spot mosaic virus. One of the most significant abiotic stresses is salt stress. Increases in salinity tolerance in rice is may be attributed to the transcription factor *OsRR22* (Isbel *et al.* 2022).

Resistance to Disease

To cause infection, viruses infiltrate their host's metabolic processes. Since they facilitate viral RNA translation and infection, host components involved in translation have been labeled as pro-viral vectors (Kusnadi *et al.* 2022). The efficient manifestation of the viral infection depends on a wide variety of other host factors as well. It has proven possible to employ genes related to these factors to control infectious diseases in plants. These variables are involved in translation, replication, metabolism, and other activities.

CRISPR/Cas9's Recent Advances in Plant Viral Disease Management

Nucleotide insertions and deletions at certain sites disrupt genes by introducing mutations that interfere with their normal function. The promoter and the coding region of a gene are both potential sites for the introduction of Indels (insertions and deletions). The detrimental effects of viruses on plants may be modified by modifying the activation of disease susceptibility genes in the plant (NISA *et al.* 2022). Using a homology directed repair (HDR) mechanism, the CRISPR approach may also be used to replace nonfunctional resistance (*R*) genes in plants with a functional *R* gene from a virus-resistant plant variety.

By editing specific regions of the virus's genome, the CRISPR-Cas9 system is utilized to develop creatures that are resistant to viral infections. However, CRISPR-Cas9-mediated viral resistance has been mostly achieved in geminiviruses with monopartite or bipartite genomes by ssDNA targeting (Chauhan *et al.* 2022). Included in its DNA is the coding for proteins necessary for viral replication, migration and defensive evasion.

Guided RNA was often positioned on stem-loop sequences in the intergenic region close to the replication origin, and this was shown to be the case in the vast majority of functional targeting. To some extent, protection against other geminiviruses, such the bipartite MeMV (Merremia mosaic virus) and the monopartite beet curly top virus (BCTV), may be afforded by the intergenic area replication origin shared by almost all geminiviruses (beet curly top virus) (Karmakar *et al.* 2022). According to the study, targeting the noncoding intergenic region leads to long-term resistance by limiting the development of viral variants with the potential to migrate and multiply. This study is a beacon for the scientific community since it takes a genome-level approach to combating viruses, which has shown effective in the past (Akram *et al.* 2022).

The studies mentioned above also show that the CRISPR-Cas9 method is effective in creating CLCuKoV-resistant cotton (Cotton leaf-curl Kokhran virus). When first isolated from *Streptococcus pyogenes*, Cas9 could only alter dsDNA; therefore, it was used only to target DNA viruses. *Francisella novicida* (FnCas9) was found to do DNA modifications, while *Leptotrichia wadei* (LwaCas13a), previously known as C2c2, was discovered for the first time for targeting the RNA viral genome directly(Saeed *et al.* 2022). The applications of sgRNA and FnCas9 in *Arabidopsis thaliana* and *Nicotiana benthamiana* significantly reduced viral accumulation and slowed the progression of CMV (cucumber mosaic virus) and TMV (tobacco mosaic virus) symptoms, respectively.

N. benthamiana was targeted with the CRISPR-Cas13a system to repress the RNA virus TuMV (turnip mosaic virus). Compared to CP (coat protein) targeting, GFP sequences with HC-Pro targeting increased resistance. The conversion of poly-crRNA into single crRNAs has been effectively characterized by the inheritability of Cas13a. Despite the fact that Cas13b, a member of the Cas13 co-family, is more resistant to RNA knockdown than Cas13a, very little is known about its use in plants

(Bansal *et al.* 2022). The host susceptibility factor, the eukaryotic translation initiation factors eIF(iso)4E, and 4E, became the most widely used strategy against RNA viruses when an RNA-guided editing strategy was not discovered (*eIF*4E). The *elF*4E gene is the largest group of genes that produce recessive viral resistance and has been found in both monocots and dicots as a critical host susceptibility component during viral infections. On the other hand, *eIF* (iso)4E and *eIF*4E showed recessive resistance to viral infections in melons and tomatoes (Milazzo *et al.* 2022).

CRISPR has the potential to replace more traditional methods of plant breeding, transgenes, and genome editing. One possible use of CRISPR is in the treatment of chronic diseases. The need for rapidly induced resistance is exacerbated by the fact that the plants are perennial and hence rooted firmly in one place. These two strategies are virus-resistant and may find use in today's farming. However, we need to develop new methods of insuring dependability and implementing new mechanisms of action (Malik *et al.* 2022). It is possible to create virus-resistant plants by targeting both the coding and noncoding regions of the viral genome. Mutations in noncoding regions have shown to be highly effective in the evolution of infection-resistant plants, in part, because viral coding sequences adapt swiftly to deal with plant defensive systems. To create plants with higher levels of viral resistance, researchers used gRNA-Cas9 nuclease proteins to target noncoding regions such as capsid protein, replication proteins and so on (Shakir *et al.* 2022). The noncoding sections of viruses like Merremia mosaic virus (MeMV), tomato yellow leaf-curl virus (TYLCV) and beet curly top virus (BCTV) are the focus of mutagenesis research. These results show that the CRISPR-Cas system is an effective and adequate tool for generating targeted mutagenesis (Angon and Habiba, 2023).

The gRNA might be altered to stop the virus from producing resistant strains in the host plants. Viruses that consist only of RNA are called RNA viruses. More effective against RNA viruses are variations of the Cas protein, such as *F. novicida*'s FnCas9 and *L. wadei*'s CRISPR-Cas effector LwaCas13a. Using FnCas9-gRNA technology, we are able to lessen the severity of infection and the number of viruses in the body (Shahriar *et al.* 2021) (Table 3.1).

Fungal Resistance Development Using CRISPR-Cas9 Technology

Fungal resistance has been mainly engineered with CRISPR-Cas9 by editing S genes, including mildew resistance locus O (MLO), rice Ethylene Response Factor 922 and others. Commonly known *S* gene locus, O confers resistance to mildew (MLO). Barley resistant to powdery mildew was first discovered in 1942, and since then various varieties have been developed (Liu *et al.* 2022). Seven transmembrane domains and a calmodulin-binding domain characterize the structure of MLO—a protein found in the plasma membrane. Powdery mildew susceptibility in monocots and dicots has also been linked to it. By using TALEN and the TaMLO-A1 allele of exon 2 of bread wheat, Wang *et al.* (2014) targeted *MLO*-A1, *MLO*-B1, and *MLO*-D1 homoeoalleles using CRISPR-Cas9 technology. Heritable resistance to powdery mildew caused by *Blumeria graminis* f. sp. *tritici* was established using both methods. Tomato powdery mildew is brought on by the Ascomycete fungus *Oidium neolycopersici* (Tateno *et al.* 2022).

TABLE 3.1
CRISPR-Cas9 Technology Applications for Creating Resistance in Plants against Viral Disease

Resistant against Viral Disease	Target Region	Induced Function: Halting Rolling Circle Replication	Induced Function: Inhibition of Viral Replication	Induced Function: Host Resistance	Targeted Plant	Reference
BSCTV	CP, Rep and IR	✓	✓	✓	*Nicotiana benthamiana* and *Arabidopsis thaliana*	Ji *et al.* (2015)
TYLCV, BCTV	CP, RCR II motif of Rep and IR	✓	✓	✓	*Nicotiana benthamiana*	Ali *et al.* (2015)
MeMV, TYLCV	CP, Rep and IR	✓	✓	✓	*N. benthamiana*	Ali *et al.* (2016)

To penetrate the plant cell wall and get access to the host cell, invading fungi produce enzymes. Callose is synthesized by plants to fortify their cell walls against attack. Thus, inhibitors of fungal enzymes that deteriorate cell walls may be used to alter genetic material. SlPMR4, an ortholog of Powdery Mildew Resistance 4, promotes calcareous deposition (PMR4) (Talakayala *et al.* 2022). A fungus called *Oidium neolycopersici* was used together with CRISPR-Cas9 to increase a plant's resilience. Improved resistance to the powdery mildew pathogen *Erysiphe cichoracearum* was achieved by inducing mutations in the wheat susceptibility gene *Taedrl* (an ortholog of EDR) using CRISPR. This gene mutation confers a wide spectrum of resistance to disease.

The bacterium *Xanthomonas oryzae* is responsible for plant blight. Resistance was achieved by modifying the *Os*SWEET13 gene, which controls the host plant's sensitivity to the disease. In order for bacteria to cause infection, this gene must be active in the transport of sucrose (Liao *et al.* 2022). Important for both ethylene signaling and resistance to pathogens, the rice genes Ethylene Response Factor 922 (ERF922) and enhanced disease resistance 1 (EDR1) are promising candidates for future genetic engineering. With the use of CRISPR-Cas9, the genes *OsE*RF922 and *Os*SEC3A in rice plants were altered, making them completely immune to the blast disease without affecting the plants development (Li *et al.* 2022) (Table 3.2).

Resistance against Bacterial Diseases Using CRISPR-Cas9 Technology

Several studies have been published on using CRISPR-Cas9 to combat bacterial infections, in comparison to viral and fungal resistance. The γ-proteobacterium Xanthomonas uses type III transcription activator-like effectors (TALEs) to increase

TABLE 3.2
Major Applications of CRISPR-Cas9 Technology for Creating Resistance in Plants against Fungal Disease

Resistance against Fungal Disease	Target Region	Induced Function: Host Susceptibility Gene	Induced Function: Transcription Factor in Multiple Stress Responses	Targeted Plant	Reference
Powdery mildew	TaMLO-A1	✓		*Triticum aestivum*	Wang *et al.* (2014)
Powdery mildew	SlMlo1	✓		*Solanum lycopersicum*	Nekrasov *et al.* (2017)
Rice blast disease	OsERF922		✓	*Oryza sativa L. japonica (var. Kuiku131)*	Wang *et al.* (2016)

host gene expression, making the host more susceptible to infection (Getahun, 2022). The sucrose transporter gene *OsS*WEET13 has been shown to be a *Xanthomonas phthoracis* susceptibility gene. The *OsS*WEET13 gene was transferred from the disease-resistant rice variety IR24 to the japonica rice variety Kitaake, which made it susceptible to the disease. However, a modification in the allele by CRISPR-Cas9 made the plant resistant to bacterial blight (Zhou *et al.* 2022).

The gene *DMR*6 is a negative regulator of plant defenses against downy mildew. During a *Pseudomonas* infection, according to de Toledo Thomazella *et al.* the DMR6 ortholog SlDMR6-1 is upregulated in tomatoes due to the presence of phytophthora (Chaudhary *et al.* 2022). Mutant plants with a truncated version of SlDMR6 that displayed broad-spectrum resistance to *Xanthomonas perforans* and *Pseudomonas syringae* were created by targeting exon3 of *SlD*MR6-1. Coronatine is produced as a result, and this facilitates bacterial colonization by opening the stomata. The COR co-receptor is essential for the stomatal response to COR in Arabidopsis. The C-terminal Jas domain is necessary for COR to open stomata; however, its absence in the truncated form of *JAZ*21jas renders the protein inactive (Deb *et al.* 2022) (Table 3.3).

Web-Based Tools and Resources Available for Designing sgRNAs

With more and more research using CRISPR-Cas methods, more and more data is being produced that might be utilized to refine computational analytic models. Multiple sgRNA designing tools have been compared, and the results demonstrate that most of them have distinguishing features. In contrast to general-purpose websites that merely provide sgRNA designing services, organism-specific resources often include CRISPR-Cas vectors and procedures that have been tested in the field.

TABLE 3.3
CRISPR-Cas9 Uses for Creating Resistance in Plants against Bacterial Disease

Resistance against Bacterial Disease	Target Region	Induced Function			Targeted Plant	Reference
		Sucrose Transporter	COR Virulence Factor	Host Susceptibility		
Bacterial blight	SWEET13	✓			*Oryza sativa*	Zhou *et al.* (2015)
Pseudomonas syringae pv.	SlJAZ2		✓		*Solanum lycopersicum*	Ortigosa *et al.* (2018)
Fire blight	DIPM-1, 2, 4			✓	*Malus domestica*	Malnoy *et al.* (2016)

CRISPOR

The technologies included inside CRISPOR range from primer design software to efficiency and specificity prediction programs to gRNA design software. These methods may be utilized for both on-target and off-target detection as well as the construction of vectors. To a high degree, these models accurately depict the predicted results. As part of its specificity prediction capabilities, CRISPOR incorporates the MIT and CFD specificity prediction tools (Blin, 2016). To further reduce cutting time, CRISPOR combines two CRISPR-Cas outcome forecasting models: out-of-frame score and frameshift ratio. Additionally, the data are annotated with descriptions of many critical metrics, such as GC content, match-vs-mismatch classification (0–4 nt) and total mismatches. In CRISPOR, data of most organisms can be accessed. Nuclease enzymes and PAMs of different types are also available for selection. Because of these features, CRISPOR is currently being used by the great majority of researchers to construct various CRISPR-Cas tools for genome editing experiments (Li *et al.* 2022). CRISPOR can be accessed at http://crispor.tefor.net/.

CHOPCHOP

CHOPCHOP is a tool that has everything you need to make sgRNA. CHOPCHOP works with both the CRISPR-Cas and TALEN tool for genome editing. CHOPCHOP also has a number of ways to target a gene, such as gene activation, gene repression, knockout and knock in. Like CRISPOR, CHOPCHOP gives a number of predictive models from which the user can choose one to make a prediction about the specificity and efficiency of the cutting site (Park *et al.* 2016). CHOPCHOP has a 'Custom PAM' feature that makes it easy to pick different PAM sequences from a drop-down menu. It has been suggested that the types of cells present may affect the DSB repair pathway, which in turn can affect the results of CRISPR-Cas genome editing. CHOPCHOP website is user-friendly to get a good prediction of the outcome (Brazelton Jr *et al.* 2015). CHOPCHOP can be accessed at https://chopchop.cbu.uib.no/.

CRISPR RGEN TOOLS

CRISPR RGEN Tool is a CRISPR-Cas library platform that contains a variety of tools for designing sgRNA. Cas-designer is used for conventional CRISPR-Cas nucleases, BE-Designer is used for CRISPR base editing and PE-Designer is used for CRISPR prime editing; all three are included into CRISPR RGEN Tools. In addition, although Cas-Designer and BE-Designer are compatible with a wide variety of PAMs, PE-Designer is only compatible with SpCas9 (Anderson *et al.* 2021). Microhomology-Predictor is an outcome-predictive tool that analyses probable in-frame deletions generated by the MMEJ repair procedure. The software includes an algorithm for rating out-of-frame areas to identify candidates for gene editing sites. Supporting CRISPR-Cas and other programmable nucleases including zinc finger nucleases (ZFNs) and transcription activator-like effector nucleases (TALENs), this tool recommends an out-of-frame score of at least 66 (Chen *et al.* 2020). CRISPR RGEN TOOLS can be accessed at http://www.rgenome.net/.

CRISPR-GE

CRISPR-GE is a web-based service for making plant-specific sgRNAs. CRISPR-GE has been applied to the genomes of 41 plant species, including some of the most economically important crops in the world, such as grapes (*Vitis vinifera*), rice (*Oryza sativa*) and maize (*Zea mays*) (Hwang *et al.* 2021). Several other Cas nucleases, including SpCas9, FnCas12a and AsCas12a, are included in this resource to aid in the creation of sgRNAs for use with different types of CRISPR-Cas systems. Furthermore, the 'User specified' setting in CRISPR-GE allows for the customization of 14PAM sequences in a number of ways, including the selection of 5′ or 3′ PAMs and the modification of the target site's length. CRISPR-GE makes use of the CFD model to provide predictions about the specificity of a target site. Furthermore, CRISPR-GE provides a primer design tool that might aid in vector construction and mutant detection (Schindele *et al.* 2020).

CRISPR-P

Similarly, CRISPR-P is a web-based tool that can be used to create sgRNAs for plants. It has 75 plant genomes, the great majority of which are those of important grain crops. When compared to CRISPR-GE, the number of CRISPR-Cas PAM types available to CRISPR-P is larger (Ma *et al.* 2014). There are a number of other CRISPR-Cas PAM types, including NGG (SpCas9), NNAGAAW (StCas9), N4GMTT (NmCas9), NNGRRT (SaCas9), and NG. In CRISPR-GE, NNAGAAW (StCas9) is the sole RNA editing tool available (xCas9). When making new sgRNAs, CRISPR-P users may choose between using the U3 or U6 sgRNA promoter-driven expression cassette. Detailed information about the sgRNA prediction, such as the GC content, restriction endonuclease site location, secondary structure and microhomology score may be shown (Yu *et al.* 2017).

SNP-CRISPR

Using public variant data sets or user-identified variations, SNP-CRISPR, a web-based computational application, may construct sgRNAs. Particularly useful for SNP-containing alleles, it may be applied to a wide range of genetic backgrounds and model animals (Richter *et al.* 2013).

PnB Designer

PnB Designer is a web-based application that may be used to build sgRNAs for both prime and base editors, which are two recently developed CRISPR-Cas genome editors. PnB Designer is capable of designing sgRNAs for single as well as multiple genome targets, and it can apply these designs to a wide variety of plant and animal species (Fuster-García *et al.* 2017).

Sequence Scan for CRISPR (SSC)

SSC, which may be accessed online at http://cistrome.org/SSC/, is a website that scans sgRNA spacers. Not only can it be used for the design of sgRNAs for CRISPR knockout, but it can also be used for the inhibition or activation of CRISPR via the prediction of sgRNA effectiveness (Chen *et al.* 2017). In addition to computational tools built by academic institutions and made accessible to the public, certain CRISPR firms have also established a number of helpful computational tools and resources for the general public (Table 3.4).

Developments in Delivery of Gene-Editing Reagents into Plant Cells

Gene editing relies on plant genetic transformation and regeneration as two of its foundational mechanisms. On the other hand, these processes constitute significant

TABLE 3.4
Web-based Tools for gRNA Designing

Name	Organism	Cas Nuclease Enzyme	Database Web Servers	Website	References
CRISPROR	>100	>30 (Cas9 orthologs and Cas variants)	Web Server	http://crispor.tefor.net/	Haeussler *et al.* (2016)
CHOPCHOP	>100	Cas9, Cas12, Cas13 and TALEN	Web Server	https://chopchop.cbu.uib.no/	Labun *et al.* (2016), Montague *et al.* (2014)
CRISPR RGEN Tools	> 100	> 20 (Cas9 orthologs and Cas variants)	Web server	http://www.rgenome.net/cas-designer/	Bae *et al.* (2014), Park *et al.* (2015)
CRISPRscan	> 10	Cas9 and Cas12	Web Server	https://www.crisprscan.org/	Moreno-Mateos *et al.* (2015)

bottlenecks in the populations of many species. The transport of genetic material to plants via the medium of agrobacterium is by far the most prevalent method of plant genetic transformation (Atkins & Voytas, 2020). In this method, the DNA that has to be transferred is first inserted into the plant's transfer T-DNA, and, after that, the transfer DNA is eventually incorporated into the plant's genome. The second approach that is often used when dealing with monocot species is called particle bombardment, and it makes use of a gene cannon. Both the methods introduce an element of random chance into the process by inserting DNA into plant genomes. Any process in which foreign DNA is introduced into the genome of an organism is considered a kind of genetic alteration and requires the oversight of regulatory bodies (Nadakuduti *et al.* 2021). When compared to the cell walls of other kinds of cells, the cell walls that are present in plant cells provide an especially challenging barrier for the transfer of chemicals that are utilized in gene editing. It is not necessary to use transgenesis in order to edit the genome using protoplasts, which are cells that resemble animal cells but only have plasma membranes. Protoplasts present a possibility for doing so.

Gene Editing by Expression of Developmental Regulators and De Novo Meristem Induction in Plants

When subjected to transitory expression, developmental regulators (DRs) such as BABYBOOM (BBM) and WUSCHEL (WUS) have been shown to promote somatic embryogenesis in plants. This somatic embryogenesis ultimately results in the genetic transformation of lines that were previously resistant to such changes. Agrobacterium injection is one technique that has been used for gene editing. This technique involves inducing meristems in somatic cells by ectopically expressing DRs such as BBM, WUS, Shoot Meristemless (STM) and Isopentenyl Transferase (IPT) (Maher *et al.* 2020). By transiently introducing guide RNAs and DRs to *N. benthamiana* plants that were overexpressing Cas9, the researchers were able to edit genes in a way that ensured the changes would be inherited. Either by co-culturing seedlings that sprouted in liquid culture with agrobacterium or by injecting agrobacterium into plants that were grown in soil, this goal may be accomplished (Nasti & Voytas, 2021). Cas9 overexpression occurred in the *N. benthamiana* plants as a consequence of both of the techniques. When transgenic plants manufacture Cas9 on their own all the time, gene editing that makes use of the de novo meristem induction strategy may be considered a technique with relatively high throughput for the aim of editing genes. This is largely attributable to the fact that time-consuming activities associated with tissue culture may be omitted (Nadakuduti & Enciso-Rodríguez, 2021).

RNA Viruses and Mobile Guide RNAs for Heritable Plant Gene Editing

Using a positive-strand RNA virus, such the tobacco rattle virus (TRV), to deliver sgRNAs into Cas9 overexpressing plants through agrobacterium infiltration is a high-throughput heritable gene-editing approach. This is a high-throughput approach. To achieve systemic gene editing with heritable mutations, sgRNAs are coupled with

RNA mobile elements such as the Flowering locus T (FT) to increase reagent mobility to apical meristems, creating germ line alterations. Systemic gene editing is possible (Ellison *et al.* 2020). Agrobacterium is utilized to infiltrate plants with the TRV vector containing the modified sgRNAs. Nadakuduti and Enciso-Advances Rodriguez's Genome Editing by CRISPR Systems and Delivery Methods in Plants shows that this method creates heritable bi-allelic alterations without viral transmission to progeny. Due of the low cargo capacity of positive strand RNA or DNA viruses, this approach cannot deliver entire CRISPR-Cas9 expression cassettes to plants (Lei *et al.* 2021).

Nanoparticles for Delivering Biomolecules to Facilitate Plant Genome Engineering

Nano carriers provide a one-of-a-kind chance to carry biomolecules into plants while protecting them from destruction inside plant cells; agriculture is only starting to explore the promise of nanotechnology. If one of a substance's dimensions is 100 nm or less, called as nanoscale. The size exclusion limit of plant cells' hydrophilic cell walls is between 5 and 20 nm, whereas that of the lipid plasma membrane, which lines the cell's interior, is 500 nm (Cunningham *et al.* 2018).

In the biolistic transformation method, heavy metal nanoparticles (NP) are employed to deliver their payload through a gene gun. Carbon dots with a diameter of less than three nanometers and single-walled carbon nanotubes (CNTs) with a diameter between 1 and 1,000 nm are capable of being chemically functionalized to carry genetic material, can diffuse through the walls of plant cells and can deliver cargo to specific cell organelles (Wang *et al.* 2019). In order to silence genes, scientists have recently shown that carbon nanotubes (CNTs) and carbon dots may efficiently transfer DNA into nuclear and chloroplast genomes. Without introducing any foreign biolistics or chemicals, and without inserting any DNA into fully developed plants, this was achieved. For the purpose of transporting biomolecules (DNA/RNA/proteins and RNPs) into plant cells for the purpose of affecting either the germline or somatic tissues, nano-carbons such as carbon nanotubes (CNTs), fullerenes, graphene and polymeric nanoparticles (NPs), especially polyethyleneimine-coated NPs, show great promise (Wang *et al.* 2019).

PEG-Mediated CRISPR-Cas9 Vector Delivery

Polyethylene glycol (PEG) is used in this critical gene-editing procedure. Numerous plant protoplasts, such as those of maize, soybean, *A. thaliana*, tobacco, rice, and wheat, have been successfully cultured using this method. The protoplast is treated with the plasmid encoding Cas9 and gRNA in the presence of PEG. Both the U3 promoter for gRNA and the CaMV35S promoter for Cas9 were employed in the first report of PEG-mediated CRISPR construct delivery in maize (Sandhya *et al.* 2020). Six polyphenol oxidase mushrooms were edited without using transgenes in 2016 using ribonucleoproteins and PEG-mediated gene regulation.

The mutant mushroom avoided US regulation and had 30% less activity of the browning enzyme. The conventional floral-dip or agrobacterium-assisted methods of administration are ineffective for these ribonucleoprotein complexes. The Primary

TABLE 3.5
PEG-Mediated CRISPR-Cas9 Components Delivery into Different Plants

Plant Name	CRISPR-Ca9 Vector or Ribonucleoprotein Complexes	Targeted Genes	Reference
Apple	Cas9-sgRNA ribonucleoprotein complexes	*DIP*M-1, 2, 4	Malnoy *et al.* (2016)
Brassica oleracea, Brassica rapa	Cas9-sgRNA ribonucleoprotein complexes	*FRI, PDS*	Murovec *et al.* (2018)
Citrullus lanatus	PHSN1, PHSN2	*Cl*PDS	Tian *et al.* (2017)
Glycine max	pCas9-GmU6-sgRNA, pCas9-AtU6-sgRNA	*Gl*yma08g02290, *Gl*yma12g37050, *Gl*yma06g14180	Sun *et al.* (2015)
Grapevine	Cas9-sgRNA ribonucleoprotein complexes	*MLO*-7	Malnoy *et al.* (2016)
Oryza sativa	pRGE3, pRGE6	*OsM*PK5	Sun *et al.* (2015)

limitations and difficulties of PEG-mediated delivery are the generation of suspension cells and the isolation of protoplasts (Wu *et al.* 2020). Furthermore, resistant plant species are the largest challenge to successful protoplast-to-whole-plant regeneration. There is a need to look into other methods of delivering the ribonucleoprotein complex Cas9 and the guide RNA it uses to modify DNA (Table 3.5).

Bombardment-Mediated Delivery of Vector or Cas9/gRNA Ribonucleoproteins

In order to carry out this transformation or gene transfer procedure, a piece of equipment known as a 'gene gun' or 'biolistic gun' is necessary. Vectors or Cas9/gRNA ribonucleoproteins often use gold, silver or tungsten particles to serve as a carrier. Components of the CRISPR-Cas9 system are delivered into explants by applying high pressure, and this process requires coated particles (Sandhya *et al.* 2020). There are a number of factors that may be adjusted to improve the efficacy of this method, including helium pressure, target distance, particle size, and the explants employed. The transformed explants are regenerated onto a regeneration medium with the appropriate degree of selection pressure. Reports of successful delivery of Cas9-gRNA ribonucleoproteins and subsequent regeneration of mutants in maize, potato, and brassica have been published (Liu *et al.* 2020) (Table 3.6).

Agrobacterium-Mediated CRISPR-Cas9 Construct Delivery

This is the most common method in which agrobacterium strain is engineered to recognize and accept the Cas9 and gRNA expression cassette included inside the binary vector. Further, agrobacterium mediates the transfer of CRISPR structures into the appropriate explant, such as a plant's callus, leaf, or floral organ (Toinga-Villafuerte

TABLE 3.6
Particle Bombardment Method for CRISPR-Cas9 Component Delivery

Plant Name	CRISPR-Cas9 Vector or RNP Complex	Selectable Marker	Target Genes	Reference
Glycine max	QC810 and RTW830, QC799 and RTW831	*HptII*	*DD20*, *DD43*	Stoger *et al.* (2017)
Hordeum vulgare	pcas9:sgRNA	*HptII*	ENGase	Stoger *et al.* (2017)
Oryza sativum	pCam1300-CRISPR-B	*HptII*	*crt*I, *ZmP*sy	Dong *et al.* (2020)
Triticum aesituvam	pJIT163-Ubi	*Bar*	*TaML*O-A1, *TaML*O-B1, *TaML*O-D1	Wang *et al.* (2014)
Zea mays	pSB11-Ubi:Cas9	*Pat*	*LIG*1, *Ms*26, *Ms*45, *ALS*1, *ALS*2	Svitashev *et al.* (2015)

TABLE 3.7
Agrobacterium-Mediated Delivery of CRISPR-Cas9 Components in Different Plant Species

Plant Name	CRISPR-Cas9 Vector	Selectable Marker	Strain	Target Genes	Reference
Arabidopsis thaliana	pUC119-RCS	Marker free	GV3101	*AtP*DS3, *AtF*LS2, *RACK*1b, *RACK*1c	Li *et al.* (2013)
Arabidopsis thaliana	pCAMBIA1300	*HptII*	GV3101	*BRI*1, *GAI*, *JAZ*1	Feng *et al.* (2013)
Banana	pRGEB31	*HptII*	AGL1	*RAS*-PDS	Kaur *et al.* (2018)

et al. 2022). Agrobacterium-mediated transport of CRISPR-Cas9 components has been used to effectively modify the genomes of over 20 different plant species (Table 3.7).

When it comes to delivering gene of interest to woody plants, the agrobacterium-mediated delivery strategy is the most useful and promising. Genome editing mediated by agrobacterium has shown to be an efficient tool for changing the genetic code of woody plants like *Citrus sinensis* and poplar (Li *et al.* 2022). The plant as a whole benefits from these methods, but the easily modified tissues and organs (such leaves, callus and flowers) reap the most rewards.

Floral-Dip or Pollen-Tube Pathway Method

Plasmids were first placed to the stigmas of plants by either placing them directly on the stigma surface, or mixing them with pollen and then applying the mixture to the stigmas. Furthermore, several variables were optimized for efficient gene transfer. This was accomplished by first wounding the flowers with both male and female organs, and then immersing them into a solution of agrobacterium (Gao *et al.* 2022). The plant developmental stage is crucial to the success of its flowering metamorphosis. Critical molecular features include the selection of a promoter, the size of the gene and the types of vectors, in addition to various physical parameters like the composition of the medium, the pH level, the optical density, the temperature and the relative humidity. Moreover, the widely used constitutive promoters CaMV35S and Arabidopsis UBI10 were used into the method with the intention of improving editing efficiency. An improved form of the UBI10 promoter was found in the Arabidopsis germline (Zhou *et al.* 2022).

Pollen Magnetofection-Mediated Delivery

Magnetofection utilizes magnetic force for vector uptake in combination with magnetic nanoparticles and is a technique for genetic alteration. Magnetofection describes this method (MNP). This method employs polyethyleneimine-coated, positively-charged Fe3O4 MNPs and a negatively-charged vector to generate MNP–DNA complexes (Talakayala *et al.* 2022). After that, mix in some magnetic fields and combine the pollens and complexes. The pollens were then used in the act of pollination. The cotton sector has seen success with the use of this technology.

For the creation of nontransgenic plant life, magnetofection will be advantageous, since it does not need the use of vectors or DNA in editing. The two methods outlined below may be used to achieve this goal. First of all Cas9 mRNA and guide RNA are transcribed in vitro, and then they are coated with MNP and delivered to the stigma or protoplast (Chakravorty *et al.* 2022). When experiments are conducted in a petri dish, a T7 promoter is employed to regulate the transcription of guide RNAs and Cas9. T7 RNA polymerase may be used for in vitro transcription and DNase I can be used to clean up the final product.

There are no known instances of pollen magnetofection being employed in genome editing research at this time. One advantage is that we can insert the CRISPR-Cas9 ribo-nucleoproteins into the pollens without intervening at the DNA level (Távora *et al.* 2022). The time required for tissue culture and transgenic selection might be reduced as a result of this technique.

REFERENCES

Akram, F., Sahreen, S., Aamir, F., Malik, K., Imtiaz, M., Naseem, W., Waheed, H. M. (2022). An Insight into Modern Targeted Genome-Editing Technologies with a Special Focus on CRISPR/Cas9 and its Applications. *Molecular Biotechnology*, 1–16.

Anderson, M. V., Haldrup, J., Thomsen, E. A., Wolff, J. H., & Mikkelsen, J. G. (2021). pegIT-a web-based design tool for prime editing. *Nucleic Acids Research*, *49*(W1), W505–W509.

Angon, P. B., & Habiba, U. (2023). Application of the CRISPR/Cas9 Gene-editing System and Its Participation in Plant and Medical Science. *Current Applied Science and Technology*, 10–55003.

Atkins, P. A., & Voytas, D. F. (2020). Overcoming bottlenecks in plant gene editing. *Current Opinion in Plant Biology*, *54*, 79–84.

Bansal, S., Balamurugan, A., Achary, V., Kumar, A., Reddy, M. K., & Prakash, G. (2022). Editing plant genome with CRISPR/Cas: A sustainable strategy for disease management. In *Next-Generation Plant Breeding Approaches for Stress Resilience in Cereal Crops*. Springer Nature Singapore. pp. 369–396.

Bazaz, M. R., & Dehghani, H. J. (2022). From DNA break repair pathways to CRISPR/Cas-mediated gene knock-in methods. *Life Sciences*, 120409.

Blin, K., Pedersen, L. E., Weber, T., Lee, S., & Biotechnology, S. (2016). CRISPy-web: An online resource to design sgRNAs for CRISPR applications. *Synthetic and Systems Biotechnology*, *1*(2), 118–121.

Brazelton Jr, V. A., Zarecor, S., Wright, D. A., Wang, Y., Liu, J., Chen, K., food. (2015). A quick guide to CRISPR sgRNA design tools. *GM Crops & Food*, 6(4), 266–276.

Chakravorty, M., Nanda, M., Arora, N., Singh, S., Kumar, V., & Deshwal, S. (2022). CRISPR-Cas9 mediated genome tailoring to improve nutritional quality and shelf life in crops: A review. *Plant Gene*, 100369.

Chaudhary, M., Mukherjee, T. K., Singh, R., Gupta, M., Goyal, S., Singhal, P., Sharma, P (2022). CRISPR/Cas technology for improving nutritional values in the agricultural sector: An update. *Molecular Biology Reports*, *49*(7), 7101–7110.

Chauhan, P. K., Upadhyay, S. K., Tripathi, M., Singh, R., Krishna, D., Singh, S. K., Reviews, G. E. (2022). Understanding the salinity stress on plant and developing sustainable management strategies mediated salt-tolerant plant growth-promoting rhizobacteria and CRISPR/Cas9. *Biotechnology and Genetic Engineering Reviews*, 1–37.

Chen, C.L., Rodiger, J., Chung, V., Viswanatha, R., Mohr, S. E., Hu, Y., & Perrimon, N., Genomes, Genetics. (2020). SNP-CRISPR: A web tool for SNP-specific genome editing. *G3: Genes, Genomes, Genetics*, *10*(2), 489–494.

Chen, Y., Zeng, S., Hu, R., Wang, X., Huang, W., Liu, J.,Zhang, Y. (2017). Using local chromatin structure to improve CRISPR/Cas9 efficiency in zebrafish. *12*(8), e0182528.

Copeland, L. (2022). Resilience of agri-food systems. *MDPI*, *12*, 543.

Cunningham, F. J., Goh, N. S., Demirer, G. S., Matos, J. L., & Landry, M. P. (2018). Nanoparticle-mediated delivery towards advancing plant genetic engineering. *Trends in Biotechnology*, *36*(9), 882–897.

Deb, S., Choudhury, A., Kharbyngar, B., & Satyawada, R. R. (2022). Applications of CRISPR/Cas9 technology for modification of the plant genome. *Genetica*, 1–12.

Eissa, H. A., Mohamed, S. S., & Hussein, A. (2020). Nutritional value and impact of wheatgrass juice (Green Blood Therapy) on increasing fertility in male albino rats. *Bulletin of the National Research Centre*, *44*(1), 1–11.

Ellison, E. E., Nagalakshmi, U., Gamo, M. E., Huang, P., Dinesh-Kumar, S., & Voytas, D. F. (2020). Multiplexed heritable gene editing using RNA viruses and mobile single guide RNAs. *Nature Plants*, *6*(6), 620–624.

Fuster-García, C., García-García, G., González-Romero, E., Jaijo, T., Sequedo, M. D., Ayuso, C., Aller, E. (2017). USH2A gene editing using the CRISPR system. *Molecular Therapy-Nucleic Acids*, *8*, 529–541.

Gao, Q., Wang, C., Xi, Y., Shao, Q., Li, L., & Luan, S. (2022). A receptor–channel trio conducts Ca2+ signalling for pollen tube reception. *Nature*, *607*(7919), 534–539.

Getahun, Y. A., Ali, D. A., Taye, B. W., Alemayehu, Y. A., & Reports. (2022). Multidrug-resistant microbial therapy using antimicrobial peptides and the CRISPR/Cas9 system. *Veterinary Medicine: Research and Reports*, *13*, 173.

Hwang, G.-H., Kim, J.-S., & Bae, S. (2021). Web-based CRISPR toolkits: cas-OFFinder, cas-designer, and cas-analyzer. In *CRISPR Guide RNA Design*. Springer. pp. 23–33.

Isbel, L., Grand, R. S., & Schübeler, D. (2022). Generating specificity in genome regulation through transcription factor sensitivity to chromatin. *Nature Reviews Genetics*, 1–13.

Karmakar, S., Das, P., Panda, D., Xie, K., Baig, M. J., & Molla, K. A. (2022). A detailed landscape of CRISPR-Cas-mediated plant disease and pest management. *Plant Science*, *323*, 111376.

Kusnadi, E. P., Timpone, C., Topisirovic, I., Larsson, O., & Furic, L. (2022). Regulation of gene expression via translational buffering. *Biochimica et Biophysica Acta (BBA)-Molecular Cell Research*, *1869*(1), 119140.

Lei, J., Dai, P., Li, Y., Zhang, W., Zhou, G., Liu, C., & Liu, X. (2021). Heritable gene editing using FT mobile guide RNAs and DNA viruses. *Plant Methods*, *17*(1), 1–11.

Li, C., Chu, W., Gill, R. A., Sang, S., Shi, Y., Hu, X., Bioinformatics. (2022). Computational tools and resources for CRISPR/Cas genome editing. *Genomics, Proteomics & Bioinformatics*. https://doi.org/10.1016/j.gpb.2022.02.006.

Li, C., Zhen, S., Ma, X., Ma, L., Wang, Z., Zhang, P., Disease. (2022). Involvement of the non-canonical polyadenylation polymerase Cid14 in fungal azole resistance in the pathogen Cryptococcus neoformans. *Pathogens and Disease*, *80*(1), ftac036.

Li, G., Liu, R., Xu, R., Varshney, R. K., Ding, H., Li, M., Wang, D. (2022). Development of an Agrobacterium-mediated CRISPR/Cas9 system in pea (*Pisum sativum* L.). *The Crop Journal, 11*(1), 132–139.

Liao, B., Chen, X., Zhou, X., Zhou, Y., Shi, Y., Ye, X., Ren, B. (2022). Applications of CRISPR/Cas gene-editing technology in yeast and fungi. *Archives of Microbiology*, *204*(1), 1–14.

Lin, S.-C., Haga, K., Zeng, X.-L., & Estes, M. (2022). Generation of CRISPR–Cas9-mediated genetic knockout human intestinal tissue–derived steroid lines by lentivirus transduction and single-cell cloning. *Nature Protocols*, *17*(4), 1004–1027.

Liu, J., Vogel, A. K., Miao, J., Carnahan, J. A., Lowes, D. J., Rybak, J. M., & Peters, B. M. (2022). Rapid Hypothesis Testing in Candida albicans Clinical Isolates Using a Cloning-Free, Modular, and Recyclable System for CRISPR-Cas9 Mediated Mutant and Revertant Construction. *Microbiology Spectrum*, e02630–02621.

Liu, W., Rudis, M. R., Cheplick, M. H., Millwood, R. J., Yang, J.P., Ondzighi-Assoume, C. A., Chesnut, (2020). Lipofection-mediated genome editing using DNA-free delivery of the Cas9/gRNA ribonucleoprotein into plant cells. *Plant Cell Reports*, *39*(2), 245–257.

Ma, Y., Zhang, L., & Huang, X. (2014). Genome modification by CRISPR/Cas9. *FEBS Journal*, *281*(23), 5186–5193.

Maher, M. F., Nasti, R. A., Vollbrecht, M., Starker, C. G., Clark, M. D., & Voytas, D. F. (2020). Plant gene editing through de novo induction of meristems. *Nature Biotechnology, 38*(1), 84–89.

Malik, M. A. M., Haider, M. S., Zhai, Y., Khan, M. A. U., Pappu, H. R. (2022). Towards developing resistance to chickpea chlorotic dwarf virus through CRISPR/Cas9-mediated gene editing using multiplexed gRNAs. *Journal of Plant Diseases and Protection,* 1–11.

Mendoza, N., Losa, F., & Palacios, S. (2022). Strategy to improve the female fertility in the general gynecologist's office: Use of a nutritional supplement based on myo-inositol/D-chiro-inositol and antioxidants. *Gynecological Endocrinology*, *38*(4), 275–278.

Mianné, J., Nasri, A., Van, C. N., Bourguignon, C., Fieldès, M., Ahmed, E., Louis, A. (2022). CRISPR/Cas9-mediated gene knockout and interallelic gene conversion in human induced pluripotent stem cells using non-integrative bacteriophage-chimeric retrovirus-like particles. *BMC Biology*, *20*(1), 1–15.

Milazzo, J., Adre, H., Tharreau, D., Franco, O. L., & Meha, A. (2022). Crispr/Cas9-Targeted Knockout Of Rice Susceptibility Genes Osdja2 And Oserf104 Reveals Alternative Sources Of. *Rice Science*, *29*(6), 535–544.

Nadakuduti, S. S., & Enciso-Rodríguez, F. (2021). Advances in genome editing with CRISPR systems and transformation technologies for plant DNA manipulation. *Frontiers in Plant Science*, *11*, 637159.

Nasti, R. A., & Voytas, D. F. (2021). Attaining the promise of plant gene editing at scale. *Proceedings of the National Academy of Sciences*, *118*(22), e2004846117.

Nisa, R. T., Jan, S. K., Bhat, F. A., Rather, T. R., Nabi, A., Wani, A. A., Biology, M. (2022). Review On "Crispr-Cas9-A Genome Editing Tool For Plant Disease Management". *Plant Cell Biotechnology and Molecular Biology,* 1–14.

Panhwar, A., Abro, R., Kandhro, A., Khaskheli, A. R., Jalbani, N., Gishkori, K. A., Qaisar, S. (2022). Global Water Mapping, Requirements, and Concerns over Water Quality Shortages. *IntechOpen.*

Park, J., Kim, J.-S., & Bae, S. (2016). Cas-Database: Web-based genome-wide guide RNA library design for gene knockout screens using CRISPR-Cas9. *Bioinformatics, 32*(13), 2017–2023.

Richter, H., Randau, L., & Plagens, A. (2013). Exploiting CRISPR/Cas: Interference mechanisms and applications. *International Journal of Molecular Sciences, 14*(7), 14518–14531.

Saeed, F., Shah, T., Bukhat, S., Munsif, F., Ahmad, I., Khan, H., & Khan, A. (2022). Genome engineering as a tool for enhancing crop traits: Lessons from CRISPR/Cas9. In *Genome Editing*. Springer, pp. 3–25.

Saika, H., Mori, A., Endo, M., & Toki, S. (2019). Targeted deletion of rice retrotransposon Tos17 via CRISPR/Cas9. *Plant Cell Reports*, *38*(4), 455–458.

Sandhya, D., Jogam, P., Allini, V. R., Abbagani, S., Alok, A. J. (2020). The present and potential future methods for delivering CRISPR/Cas9 components in plants. *Journal of Genetic Engineering and Biotechnology*, *18*(1), 1–11.

Schindele, P., Wolter, F., & Puchta, H. (2020). CRISPR guide RNA design guidelines for efficient genome editing. In *RNA Tagging.* Springer, pp. 331–342.

Shahriar, S. A., Islam, M. N., Chun, C. N. W., Rahim, M. A., Paul, N. C., Uddain, J., & Siddiquee, S. (2021). Control of plant viral diseases by CRISPR/Cas9: Resistance mechanisms, strategies and challenges in food crops. *Plants*, *10*(7), 1264.

Shakir, S., Zaidi, S., Hashemi, F., Nyirakanani, C., & Vanderschuren, H. (2022). Harnessing plant viruses in the metagenomics era: From the development of infectious clones to applications. *Trends in Plant Science,* 297–311.

Sukegawa, S., Saika, H., & Toki, S. (2021). Plant genome editing: Ever more precise and wide reaching. *The Plant Journal*, 106(5), 1208–1218.

Talakayala, A., Ankanagari, S., & Garladinne, M. (2022). CRISPR-Cas genome editing system: A versatile tool for developing disease resistant crops. *Plant Stress*, 100056.

Tateno, M., Umeyama, T., Inukai, T., Takatsuka, S., Hoshino, Y., Yamagoe, S., Miyazaki, Y. J. (2022). Examination of Cyp51A-mediated azole resistance in Aspergillus lentulus using CRISPR/Cas9 genome editing. *Medical Mycology Journal*, *63*(2), 27–35.

Távora, F. T., dos Santos Diniz, F., de Moraes Rêgo-Machado, C., Freitas, N. C., Arraes, F. B. M., de Andrade, E. C., Biotechnology. (2022). CRISPR/Cas-and topical RNAi-based technologies for crop management and improvement: Reviewing the risk assessment and challenges towards a more sustainable agriculture. *Frontiers in Bioengineering and Biotechnology,* 10.

Teske, S., & Nagrath, K. (2022). Global sector-specific Scope 1, 2, and 3 analyses for setting net-zero targets: Agriculture, forestry, and processing harvested products. *SN Applied Sciences*, *4*(8), 1–19.

Toinga-Villafuerte, S., Janga, M. R., Isabel Vales, M., & Rathore, K. S. (2022). Green fluorescent protein gene as a tool to examine the efficacy of Agrobacterium-delivered CRISPR/Cas9 reagents to generate targeted mutations in the potato genome. *Plant Cell, Tissue and Organ Culture (PCTOC),* 1–12.

Tomlinson, I. (2013). Doubling food production to feed the 9 billion: A critical perspective on a key discourse of food security in the UK. *Journal of Rural Studies*, *29*, 81–90.

Wang, J. W., Grandio, E. G., Newkirk, G. M., Demirer, G. S., Butrus, S., Giraldo, J. P., & Landry, M. P. (2019). Nanoparticle-mediated genetic engineering of plants. *Molecular Plant*, *12*(8), 1037–1040.

Wang, P., Zhao, F.J., & Kopittke, P. M. (2019). Engineering crops without genome integration using nanotechnology. *Trends in Plant Science*, *24*(7), 574–577.

Wang, X. (2022). Managing land carrying capacity: Key to achieving sustainable production systems for food security. *Land, 11*(4), 484.

Wu, S., Zhu, H., Liu, J., Yang, Q., Shao, X., Bi, F., Yi, G. (2020). Establishment of a PEG-mediated protoplast transformation system based on DNA and CRISPR/Cas9 ribonucleoprotein complexes for banana. *BMC Plant Biology*, *20*(1), 1–10.

Wudil, A. H., Usman, M., Rosak-Szyrocka, J., Pilař, L., Boye, M. (2022). Reversing years for global food security: A review of the food security situation in sub-Saharan Africa (SSA). *International Journal of Environmental Research and Public Health, 19*(22), 14836.

Xia, S., Ventura, I. M., Blaha, A., Sgromo, A., Han, S., Izaurralde, E., evolution. (2022). Rapid gene evolution in an ancient post-transcriptional and translational regulatory system compensates for meiotic X chromosomal inactivation. *Molecular Biology and Evolution*, *39*(1), msab296.

Yu, J., Baek, K., Jin, E., & Bae, S. (2017). DNA-free genome editing of Chlamydomonas reinhardtii using CRISPR and subsequent mutant analysis. *Bio-protocol*, 711, e2352.

Zhou, M., Luo, J., Xiao, D., Wang, A., He, L., Zhan, J. (2022). An efficient method for the production of transgenic peanut plants by pollen tube transformation mediated by Agrobacterium tumefaciens. *Plant Cell, Tissue and Organ Culture (PCTOC)*, 1–8.

Zhou, Y., Xu, S., Jiang, N., Zhao, X., Bai, Z., Liu, J., Lv, C. (2022). Engineering of rice varieties with enhanced resistances to both blast and bacterial blight diseases via CRISPR/Cas9. *Plant Biotechnology Journal*, *20*(5), 876–885.

4 Methods for Designing Disease-Resistant Plants Using CRISPR Technology

INTRODUCTION

The main and most pressing danger to agricultural progress and global food security is plant diseases. According to data released by FAO in 2015, several diseases are responsible for 20%–50% of crop losses globally (Pests, 2015). Currently, one of the important methods for preventing crop diseases is the use of chemical pesticides. The use of pesticides, however, poses a direct or indirect risk to other living things (Damalas and Eleftherohorinos, 2011). Due to their frequent and erratic usage, phytopathogens are also becoming increasingly resistant to many pesticides and other chemicals (Ahmad *et al.* 2019). Therefore, it is crucial to avoid using pesticides in food production to reduce their detrimental effects on the ecosystem and conserve animals and flora. However, a successful and environmental approach to sustainable agriculture is the production of disease-resistant crops through selective plant breeding. In the past few decades, traditional breeding of disease-resistant cultivars has been effective, but it has many drawbacks, including the introduction of resistance (R) genes and labor intensiveness.

The most severe restrictions on conventional breeding for resistance are crossing or selfing between the two most suitable plants, a lack of genetic diversity in the plant population and the transfer of unwanted genes or characteristics together with desired resistance genes or characters (Gao, 2018). Therefore, keeping up with disease development and rising food demand is a major issue for traditional breeding, especially in the face of global climate change. Although mutant breeding and transgenic technologies are already in use, they also have several drawbacks that make them less demanding than they formerly were. These issues with our existing agricultural methods point to the necessity of introducing better and more efficient ways (such as genome-editing techniques or GETs) of producing resources for breeding novel crops.

CRISPR is one of the latest GETs that is often used to create appropriate plant material for long-term food production (Zaidi *et al.* 2019). Contrary to traditional mutagenesis techniques, the CRISPR-associated protein 9 (Cas9) systems can produce effective and transgene-free genetic manipulation in plants through a variety of methods, including protoplast transformation and a direct bombardment of Cas9 and gRNA, prokaryotes (Liang *et al.* 2018), transient expression of CRISPR-Cas9 ingredients supplied at the callus step (Zhang *et al.* 2016) and CRISPR-Cas9-derived

DOI: 10.1201/b22901-4

cytoplasm. However, compared to standard *Agrobacterium tumefaciens* T-DNA transformations, Gemini viral genome replicon can increase gene-targeting frequency approximately upto two folds. Virus-induced genome editing (VIGE), a virus-based gRNA administration method for CRISPR-Cas9-mediated plant genetic modification, has been utilized as a successful tool for genome editing (Yin *et al.* 2015). In this approach, gRNAs might be delivered into transgenic crops expressing Cas9 to enable systemic genetic modifications.

But these developments have substantially increased the CRISPR toolkit's capabilities. In the nut shell genome-editing techniques, notably CRISPR-Cas9 can spectacularly improve plant disease resistance. Ali *et al* reported that Crispr Cas 9 genome editing tools can be implicated in targeting Tobacco rattle virus, an RNA virus that multiplies in the cytoplasm, and cabbage leaf curl virus a DNA virus that multiplies in the nucleus for more robust crop production (Ali *et al.* 2015).

Site-directed genome changes are made possible by gene-specific endonuclease-based systems that generate DSBs in desired genes with a relatively low chance of off-target consequences. Crop improvement has largely relied on four separate site-specific endonuclease-based mechanisms, including CRISPR-related Cas9 protein, ZNFs, TALENs and Meganucleases. To facilitate the desired alterations of the nucleotide sequence at the specific site, specially modified nucleases are often designed to catalyze DSBs at a specific position in the sequence. The sliced DNA is then repaired via the cell's internal DNA repair mechanism HDR and NHEJ pathways. DNA break repair via NHEJ entails uniting the two sides of a double-strand break (DSB), which frequently results in random indels of different lengths and may result in frameshift mutations. HDR required exogenously given DNA sequences that are identical to DSBs to accurately direct DSB repair. A single-base genetic muations on large scale can be induced by HDR, based on the externally supplied DNA characteristics. Thus, precise genetic manipulation or specific target GE made possible by NHEJ or HR may lead to the development of unique plant varieties mostly with the incorporation of beneficial features for agronomy or the removal of undesirable features.

GENOME EDITING

Meganucleases

Meganucleases are developed from natural sources of enzymes that are transcribed by transportable introns. Meganucleases are known as designed homing endonucleases (Pâques and Duchateau, 2007; Smith *et al.* 2006; Voytas, 2013). These proteins may be designed to identify novel DNA target regions and are endogenous mediators of genomic targeting. The effective meganucleases for use in genetic manipulation are the yeast I-SceI Meganucleases, which are transcribed by an intron inside the mitochondrial ribosomal RNA (Pâques and Duchateau, 2007; Voytas, 2013). Meganucleases normally function as a duplex of two similar subunits and carry substantial DNA sequences that recognize fingerprints from 20 to 40 bp (Voytas, 2013). Despite being very tiny (165 aa for a meganuclease monomer), it is difficult to modulate their target specificity, and hence they are currently less often utilized than other SSNs.

ZINC FINGER NUCLEASES

Zinc finger nucleases (ZFNs) are specialized DNA binding proteins that may be targeted to cleave DNA sequences at predetermined locations (Carroll, 2011). ZFNs enable targeted genetic manipulation by inducing DNA double-strand breaks (DSBs) that allow homologous recombination to alter the genes. Each ZFN has a DNA-binding region that identifies a specific 6-bp hexamer in the nucleotide sequence as well as a DNA-cleaving motif made up of a FokI nuclease region that can cut DNA (Carlson *et al.* 2012). A zinc finger protein (ZFP) is created by combining these domains. A highly specialized genomic scissor is created when the DNA-binding and DNA-cleaving domains are combined (Carlson *et al.* 2012; Gupta *et al.* 2012). Site-specific DSBs are added to the DNA sequence by ZNF-mediated gene targeting, which also ligates the DSBs to significantly alter the genome (Carlson *et al.* 2012; Gupta *et al.* 2012). The creation of ZFPs with the ability to precisely target a particular DNA sequence inside the genome is what makes the use of ZFN-based genetic modification so crucial.

For creating appropriate ZFNs with the necessary sequence specificities, the Cys2His2ZFP offers the best structure conceivable (Pabo *et al.* 2001). This ZFP has a structure that is maintained by the chelate of zinc ions to persistent Cys2His2 amino acids and contains around 30 amino acids (Thakore and Gersbach, 2015). By integrating its α-helix into the main groove of such DNA double helix, the ZF motif attaches to the DNA sequence in the genomes (Pavletich and Pabo, 1991). The associations of ZFN with the nucleotide sequence are generally sequence specific and are caused by the amino acids at locations -1, +1, +2, +3, +4, +5 and +6 of the α-helix of the zinc finger (Pavletich and Pabo, 1991). A DNA triplet motif is bound by each finger. By joining many zinc finger motifs to produce ZFPs, it is feasible to bind to a longer DNA strand (Liu *et al.* 1997). A ZFN or ZFP for gene modification is created by joining ZFP any of the effector domains; methylase region (M), FokI-cleavage region (N), transcription activator site (A) and transcription repression region (R).

TRANSCRIPTION-ACTIVATOR-LIKE EFFECTOR NUCLEASES

As a substitute for ZFNs as methods for efficient genetic manipulation in plants, TALENs have been developed (Alwin *et al.* 2005). In theory, TALENs employ DSBs similarly to how ZFNs do. ZFNs and TALENs both have nonspecific FokI endonucleases. FokI regions are fused with certain DNA-binding regions of conserved sequences generated by TALEs or ZFPs (Joung and Sander, 2013). Xanthomonas bacteria produce TALE proteins, which are lease to change the transcription of genes in their host plants (Boch and Bonas, 2010; Bonas *et al.* 1989). TALEs have up to 30 replicas of highly conserved 33–34 A.A region in DNA, with the exception of the 12th and 13th positions

The 12th and 13th amino acids are known as the Repeat-Variable Di-residues (RVD), and they have a strong association with the identification of certain nucleotides. Owing to this property, as RVD can distinguish every single nucleotide, it can be utilized to generate additional DNA binding sites in any particular genomic sequence. The FokI domain operates as a dimer, and a hybrid nuclease may be created using the FokI endonuclease's nonspecific DNA cleavage motif (Bitinaite *et al.* 1998; Mani *et al.* 2005; Wah *et al.* 1998). Important factors that determine the activity of TALENs include the distance among bases between two distinct TALEN-binding sites and the number of residues of amino acids between the DNA-binding region and the FokI cleavage domain.

CRISPR-Cas9 NUCLEASES

CRISPR-Cas was discovered in bacteria. It is a part of adaptive immunity provides significant protection especially against DNA viruses and plasmids. CRISPR-Cas9 editing is a potent tool for modifying plant agronomic characteristics (Mohanta *et al.* 2017). CRISPR-Cas9 (SpCas9) has swiftly become a vital tool in several domains of agricultural research fields along with many other scientific fields. In the CRISPR-Cas9 mechanism, a sgRNA may attach to Cas9 and target specific DNA regions (Ding *et al.*, 2018). The PAM region restricts the available nucleotide sequence in a targeted gene. The Cas9 nuclease, the PAM, the CRISPR RNA (crRNA) and the trans-activating CRISPR RNA (tracrRNA) make up the CRISPR system. The exogenous nucleotide sequence is incorporated into the CRISPR cluster by naturally existing CRISPR systems (Sander and Joung, 2014). The foreign DNA-containing CRISPR cluster then generates a crRNA that is around 40 nt long and contains the PAM domain, which is complementary to the location of the gene. A guide RNA (gRNA) is created when the crRNA and tracer RNA hybridize. The gRNA connects to Cas9 and triggers the Cas9 system.

The complementary sequence pair between the targeted DNA and the Cas9 nuclease is directed by 20 nucleotides at the 5' end of the gRNA, resulting in RNA–DNA complementary base pairing (Sander and Joung, 2014). The occurrence of a PAM sequence downstream of the targeted gene, which typically comprises 50-NGG-30 or 50-NAG-30, is a requirement for cleavage (Gasiunas *et al.* 2012). This technique allows the Cas9 nuclease function to be targeted to any DNA sequence (Sander and Joung, 2014). The Cas9 system causes DSBs, which are then repaired by HDR or NHEJ (Sander and Joung, 2014). Some Cas9 variants exclusively cleave at one location (nickase) of the targeting DNA's parallel or non-complementary strands. Lower amounts of NHEJ indels allow the Cas9 nickase to promote HDR (Cong *et al.* 2013; Mali *et al.* 2013). Multiple sites can be accessed and edited at once by utilizing one Cas9 nuclease and many gRNA (Liu *et al.* 2014). When one gRNA is ineffective at disrupting a particular gene or when many genes need to be changed at once, this approach is highly helpful.

METHODS OF GENOME EDITING

Genome Editing by CRISPR-Cpf1

Recent discoveries of CRISPR-Cas systems by single-effector nucleases, such as CRISPR of *Prevotella* species *Francisella* 1 (Cpf1), C2c1, C2c2, and C2c3 effector nucleases, have varied the genome-editing toolkit (Liu *et al.* 2017). Similar to CRISPR-Cas9 but with unique characteristics that set it apart, CRISPR-Cpf1 is an

RNA-guided class II CRISPR-Cas system (Liu *et al.* 2017). Despite the existence of many Cas9 orthologs with fewer genes, like SaCas9 (Ran *et al.* 2015) as well as CjCas9 (Kim *et al.* 2017), Cpf1 is usually smaller than other Cas9 orthologs that contain a PAM sequence with a specific rate in the genetic makeup, which is advantageous in specifically for diagnostic purpose. Additionally, in comparison to Cas9's G-rich regions, Cpf1 depends on T-rich PAM region at the 5′ of protospacer region. Recently, synthetic Cpf1 variants increased the set of genes that Cpf1 may target (Liu *et al.* 2017). Furthermore, Cpf1 produces staggered DSB at the PAM-distal location, which may have additional benefits, especially for knockin techniques. The decreased off-target rate compared to CRISPR-Cas9 is believed to offer significant benefits in terms of safety concerns (Ran *et al.* 2015). Despite these multiple benefits, CRISPR-Cpf1 usage has remained below expectations during the previous two years. It is concluded at least partially the reason for this rate of adoption that was less than predicted is due to greater divergent indel of CRISPR Cas9. CRISPR-Cpf1 might become a more adaptable tool that may be extensively employed in a variety of experimental and practical situations after addressing the minor limitations.

RNA EDITING WITH CRISPR-Cas13

RNA-level genetic manipulation, where defective translation may be targeted to produce the required specific proteins, offers great promise for treating hereditary diseases. The controllable single-effector RNA-guided RNases Cas13 are present in class VI CRISPR systems. Using enzymatically deactivated Cas13 (dCas13) to guide adenosine to inosine deaminase function by ADAR2 to transcribing cells, type VI approaches has been analyzed to design a Cas13 ortholog competent of durable knockdown and show RNA editing. Cas13 enzymes have two Higher Eukaryotes and Prokaryotes Nucleotide-binding (HEPN) RNase domains that mediate precise RNA cleavage with a preference for targets with protospacer flanking site (PFS) motif observed biochemically in bacteria (Abudayyeh *et al.* 2016). Three Cas13 protein families have been identified to date: Cas13a, Cas13b and Cas13c (Smargon *et al.* 2017). Cas13a enzymes can be adapted as tools for nucleic acid detection as well as mammalian and plant cell RNA knockdown and transcript tracking.

CRISPR-MEDIATED *S* GENE TARGETING

Plant pathogens pose a major threat to crop productivity. Typically, phytopathogens exploit plants' susceptibility (*S*) genes to facilitate their proliferation. Disrupting these *S* genes may interfere with the compatibility between the host and the pathogens, and consequently provides broad-spectrum and durable disease resistance. The production of crops is seriously threatened by plant diseases. Several commercially significant species of plants acquire wide-ranging disease resistance when an *S* gene is disrupted. In numerous commercially significant plant species, interruption of the *S* gene may offers a wide range of plant resistance. The cellular infections phase of potyviruses, single-stranded positive-sense RNA (ssRNA+) viruses like those from the family Potyviridae, depends on the *Eukaryotic translation initiation factor 4E (eIF4E)*, also referred as a cap-binding protein. Different plant species may develop

TABLE 4.1
***S* Genes Targeted Using Genome Editing to Engineer Disease Resistance**

Sr. No.	Plant	Target Modification	*S* Gene Targeted	Targeted Pathogen/ Disease	Result	Refs
1	Arabidopsis	Gene disruption	*eIF*4E	TuMV	Resistance to TuMV	Pyott *et al.* (2016)
2	Rice	Promoter disruption	*OsS*WEET14	Bacterial blight	Increased resistance to bacterial blight	Li *et al.* (2012)
3	Tomato	Gene disruption	*DMR*6	*Pseudomonas syringae*, *Phytophthora capsici* and *Xanthomonas* spp.	Resistance against *P. syringae*, *P. capsici*, & *Xanthomonas* spp.	Li *et al.* (2012)
4	Citrus	Promoter disruption	*Cs*LOB	Citrus canker	Disease severity decreased by 83.2%–98.3%	Peng *et al.* (2017)
5	Wheat	Gene disruption	*TaM*LO	PM	Indels in target, resistance not confirmed	Shan *et al.* (2013)

immunity to potyviruses by disrupting the association between the 5′-terminal protein Viral Protein Genome-Linked (VPg) in potyviruses and *eIF4Es* (Bastet, Robaglia, and Gallois, 2017). Causing disruption in *eIF4Es* with the CRISPR system has been linked to Nipomo-resistant strains in Arabidopsis and Cucumber, according to two independent investigations. Furthermore, none of the above mentioned genome-edited crops included the transgene. As shown in Table 4.1, the Resistant (*R*) gene, is the gene encodes an immune receptor that permits the identification of pathogen-derived pathogenicity product (typically an effector) and activates host defense.

The capacity to create accurate base edits far beyond four transitional mutations has recently been a fundamental constraint of the present base-editing tools, despite the extensive editing capacities of such cytosine base editor and the adenine base editor. In 2019, it was suggested that prime editing technique may be used to address these drawbacks (Anzalone *et al.* 2019). Prime editing does not need DSBs, similar to CRISPR-mediated base editing. Prime editors utilize a prime-editing guide RNA (pegRNA) and a designed reverse transcriptase coupled to Cas9 nickase. The peg RNA, which varies greatly from conventional sgRNAs and is crucial to the operation of the system, is an important point. The pegRNA additionally carries a second sequence that spells the intended sequence alterations in contrast to (a) the analogous sequence to the targeted cells that drive nCas9 to its target gene (Anzalone *et al.* 2019). The primer binding site's (PBS) entire region on DNA is bound by the 50 nt of the pegRNA and exposes the non-complementary strand. Cas9 nicks the unattached DNA of the PAM-containing strand to provide a primer for the nCas9-linked reverse transcriptase (RT). The target area is then modified in a programmed way as a result of the RT's extension of the nicked PAM strand utilizing the pegRNA's interior as a template. This procedure yields two overlapping PAM DNA folds, the original, untreated 50 folds and the modified 30 flaps that were reverse transcribed from the pegRNA.

The completely complementary strands would probably be thermodynamically preferred in the equilibrium mechanism that determines which flap hybridizes with DNA strand devoid of PAM. However, during lagging-strand DNA synthesis, the 50 flaps are primarily destroyed by cellular endonucleases (Hosfield *et al.* 1998). Finally, via cellular replication and repair processes, the hetero duplex that results, consisting of the unedited strand and the altered 30 flaps, is resolved and successfully incorporated into the host genome. The initial generation of PEs (PE1) consisted of M-MLV RT coupled to the c-terminus of nCas9 and pegRNA produced on a different plasmid. The editing efficiency of PE1 achieved a high of 0.7%–5.5% (Anzalone *et al.* 2019). Anzalone and colleagues investigated various M-MLVRT variants that were found to improve binding, catalytic processivity and thermos stability in order to further increase the effectiveness of the reverse transcriptase.

Finally, the researchers directed a different sgRNA to generate a nick on the non-edited strand, directing DNA repair to that strand using the edited strand as a template, as was previously utilized to improve editing in CBE and ABE methods. This resulted in the creation of the most recent prime editor known as PE3, which carried out all 24 single-nucleotide alterations (24 transition and transversion mutations) with an actual editing efficiency of 33% (7.9%). (Anzalone *et al.* 2019). As a result of the requirement for complementarity at Cas9 binding, PBS binding and RT product complementarity for flap resolution, the frequency of off-target effects reported with PEs was significantly reduced (Anzalone *et al.* 2019).

Prime editing has additional benefits over earlier CRISPR-mediated base-editing methods, such as no 'bystander' altering and less onerous PAM requirements because of the variable length of RT sequence. Despite this, prime editing is currently inefficient, and further research is needed to determine its specificity and the possibility of off-target alterations. The most recent base-editor generations offer improved efficiency rates and lower-indel creation, and they are considerably better described, particularly in vivo. Therefore, wherever feasible, these tools should be utilized instead of primary editors at this time. In any event, the prime-editing system represents a significant advancement toward the creation of an all-encompassing technique for genome editing, and its potential for use in treating known harmful mutations might be substantial.

DNA BASE EDITING

Two essential elements make up DNA base editors: The Cas enzyme for programmed DNA-binding and the single-stranded DNA modifying enzyme for targeted

TABLE 4.2
Genetic Payload of Base-Editing Tools

Sr. No.	CRISPR-Tool	Function	Gene Size (kb)
1	SaCas9	Nuclease	3.2
2	CjCas9	Nuclease	2.9
3	xCas9	Nuclease	3.7
4	UGI	Inhibits UNG	0.3
5	BE4	Cytosine Base editor	5.6

nucleotide modification. Cytosine base editors and adenine base editors are the two kinds of DNA base editors that have been reported frequently. With the available CRISPR-Cas BEs, all four transition mutations (C.G to T.A and A.T to G.C) may be inserted simultaneously. The development of two innovative base-editor designs that effectively cause targeted C-to-G base transversion was recently described by Kurt *et al.* (2021). A dual base-editor method for multimodal editing in human cells is also reported in recent publications. Together, these novel base editors broaden the DNA base editors' ability to target transversion mutations and make it possible to target more complicated compound modifications with a single DNA base editor, as discussed in Table 4.2.

GENE DRIVE

By changing the likelihood that a certain allele will be transferred to off springs (instead of the Mendelian 50% probability), gene drive is a natural phenomenon and a technique in gene editing that spreads a specific set of genes throughout a population. Gene drives may be developed using various techniques. Genes can be altered, disrupted, deleted or added by using this method. Gene drives are also employed to manage fungal plant diseases. For example, a SpokI-based gene drive was created to eliminate two pathogenicity loci from the important wheat disease *Fusarium graminearum*, so this drive reduces the pathogen's pathogenicity under in vitro cultures. Gene drives might be utilized to alter weed's propensity for competition. For instance, *Rht1*, a gene primarily responsible for reduced height and escalated grain yield might also prevail in several species of weeds mainly from Poaceae family where it is implicated in altering weed's potential of growth in comparison to wheat ultimately indicating possible applications of gene drive in plants. A CRISPR drive may also be created to change the sex ratio in dioecious weeds like *Amaranthus*. The weed population is so prevented from producing pollen or ovules. Herbicide tolerance in weed populations might be overcome using the awareness campaign (Tek and Budak, 2022).

RIBO-NUCLEOPROTEINS (RNPs)

Synthetic crRNA plus tracr RNA were combined with Cas9 protein to create RNP complex. A combination of crRNA and tracr RNA was digested to create gRNA complex. It could be administered to plants cells via protoplast transformation method or via particle bombardment also on callus (Sandhya *et al.* 2020). With transformation of Cas9, RNP becomes operational and is rapidly broken down inside the cell. Due to Cas9, and RNPs' quick breakdown rates enables them to change target locations without introducing transgenes, such crops are classified being transgene-free (Wilde *et al.* 2021). Cas9 RNP transmission, however, is confined to different crops with inefficient transformation, because it requires an effective transformation method. It has also been shown that viral vectors, such as the Tobacco rattle virus (TRV), may transport CRISPR tools in germ line cells. Edits within target genome can therefore be passed down to the following generation by utilizing seedlings from plant that has its germ line altered. However, such a mechanism may have a poor level of efficiency (Wilde *et al.* 2021). Another technique uses particle bombardment to enable the temporary expression of CRISPR reagents given during the callus phase (Mushtaq *et al.* 2021). However, a number of other techniques exist and are continually enhancing transgene-free gene edited crops.

AGROBACTERIUM-MEDIATED CRISPR-Cas9 CONSTRUCT DELIVERY

Agrobacterium-mediated transformation is the most popular delivery technique for different plants. The Agrobacterium strain is used for transforming the binary vector, which contains Cas9 and the gRNA expression cassette. More than 20 different plants have been successfully altered using the CRISPR-Cas9 component delivered by Agrobacterium so far. In comparison to the particle bombardment approach, agrobacterium-mediated transformation is more effective. Agrobacterium is also widely utilized for the genetic modification of monocot crops, since they have lower capacities for regenerating and transformation. The CRISPR-Cas9 vector with good editing frequency for cotyledons and dioecious were created for Agrobacterium. The *phytoene desaturase* gene was discovered to be 59% mutated in banana genotype, Rasthali, through genome editing using Agrobacterium-mediated transformation (Kaur *et al.* 2018). Agrobacterium-mediated editing of the identical gene was found to have a 100% editing accuracy in the Cavendish banana strain 'Williams' in another study. Agrobacterium-mediated transformation is the most effective and helpful delivery mechanism, especially for woody plants. Agrobacterium-mediated genetic manipulation effectively modified woody plants like Populus and *Citrus sinensis*. This technique is often helpful for plants where leaf, callus and flowery organs may be easily transformed. A workflow of transformation and delivery methods of CRISPR is given in Figure 4.1.

Agrobacterium strain with CRISPR vector transported into protoplasts through gold particles (gene gun mediated). CRISPR constructs are transformed in the

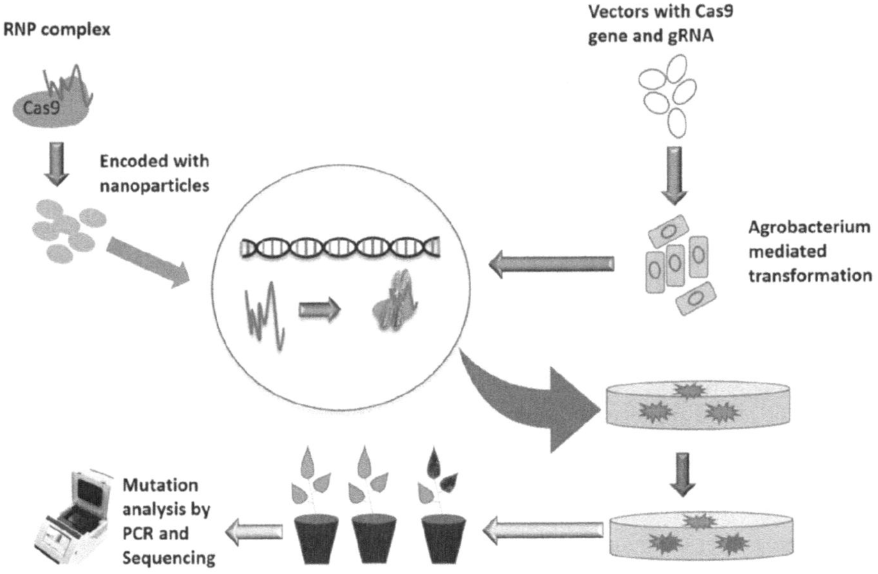

FIGURE 4.1 Transformation and delivery method of CRISPR reagents in plants.

original explants. gRNA and Cas9 are depicted in single plant tissue and at their genomic targets, respectively. The converted explants are then selected and incubated into the proper media in a plate. The plants that survived are put into containers so they could acclimatize. Mutation screening often involves PCR and Sanger sequencing.

PEG-MEDIATED CRISPR-Cas9 VECTOR DELIVERY

This crucial genetic modification technique is carried out while polyethylene glycol (PEG) is present. In the protoplasts of several plants, including maize, soybean, A. thaliana, tobacco, rice and wheat, pegRNA-mediated CRISPR has been delivered efficiently. In the presence of PEG, protoplast is cultured with the plasmid harboring Cas9 and gRNA. First described in maize, this PEG-mediated CRISPR construct delivery method utilized U3 and CaMV35S promoters for gRNA and Cas9, respectively (Liang *et al.* 2014). In a few experiments, Cas9 was induced using particular promoters that were created for certain plants and that targeted critical genes (Lowder *et al.* 2016). PEG-mediated transformation method has been used efficiently for delivering the CRISPR vector into the protoplasts. Protoplasts may be generated using an appropriate regeneration media (Lowder *et al.* 2015). In rice, Arabidopsis, tobacco and lettuce, transgene-free mutant plants were created using Cas9-gRNA ribonucleoproteins and PEG-mediated transport approximately 46% genome of lettucie has been edited via PEG-mediated CRISPR-Cas9 delivery. Later, without utilizing a vector or DNA edited the tobacco and soybean genomes using Cpf1-CrRNA ribonucleoproteins (Kim *et al.* 2017).

Cpf1, also known as CRISPR from *Prevotella* and *Francisella* 1, is a type II and type V endonuclease (Kim *et al.* 2017). It locates a PAM (TTTN) with a high thymidine content in the target area. While Cas9 requires both crRNA and tracer RNA, Cpf1-mediated genetic modification only needs crRNA. The primary benefit of PEG-mediated administration is that Cas9-gRNA ribonucleoproteins are frequently delivered using this technique. The regulatory and ethical hurdles won't likely be a problem for the vector-less or DNA-free modified mutant plants. Utilizing ribonucleoproteins and PEG-mediated technology, six polyphenol oxidase mushrooms were edited without the use of transgenes in 2016. The mutant mushroom exhibited a 30% decrease in browning-causing enzyme activity and evades US regulation (Waltz, 2016). However, the commonly employed Agrobacterium-mediated floral-dip delivery methods cannot deliver ribonucleoprotein complex in the plants. The isolation of protoplasts and the formation of suspension cells are the main difficulties and weaknesses of PEG-mediated delivery methods. In addition, some plants are reluctant to the regeneration of protoplasts into whole plant. For effective genome editing, it is necessary to investigate alternative Cas9-gRNA ribonucleoprotein delivery strategies.

KNOCKIN THROUGH THE SEQUENTIAL FLORAL-DIP METHOD

There is a need for CRISPR-Cas9 site-directed insertion of the target gene, promoter or DNA fragment at a certain spot. In the cases of tomato, maize, wheat, and potato, this has been effectively proved. In order to achieve a knockin, a donor template or donor vector that consists of the left and right homology arms is needed. Among the donor vectors, for instance, has two T-MLO homology arms as well as a GFP coding region. Together with the CRISPR-Cas9 vector, this generated GFP donor vector was introduced into wheat protoplast for GFP knock-in (Wang *et al.* 2014). The donor vector and CRISPR-Cas9 component were applied to the soybean callus. This donor DNA has a promoter unique to soybeans and the hygromycin-resistant *Hpt*II gene. With the use of the donor vector, CRISPR knock-ins into germline cells and other regenerative cells are possible. DD45, Lat52, YAO and CDC45 promoters were employed in the progressive floral-dip technique of transformation on the Cas9-carrying Arabidopsis line (Miki, *et al.* 2018). When compared to other restorative tissues, the Cas9 driven by the DD45 promoter was shown to be more effective for knock-in and to have a high rate of editing in egg cells or early embryos (Miki *et al.* 2018). Gene transfer via the floral-dip technique has been proven effective in a number of crops, including wheat (Zale, 2009), flax, radish and tomato. The Cas9 could effectively knock in targeted genes in various crops when it is activated by promoters that are specific to egg cells or embryos and the desired donor DNA template.

LIMITATIONS

Numerous research from the last 10 years has demonstrated the capability of the CRISPR system to alter the genome of a wide variety of plant species, including monocots and dicots, and to boost resistance to a variety of plant infections (Liu *et al.* 2017). However, there are two constraints present. First, infections are always changing and attempting to alter their genomes to overcome the previously present resistance. Additionally, there is a constant trade-off among plant genes, and altering the S gene may reduce the fitness of the plant. The second significant restriction is the 'off-target mutation' restriction imposed by CRISPR-Cas9 technologies.

Although it is simple to introduce one or two mutations into the host to knock out a vulnerable gene, additionally these changes are more resilient than the *R* gene mutation, because the pathogen may apply little selection pressure to counteract the plant defensive mechanism (Pavan *et al.*, 2010). However, this particular lack of mutation in the *S* gene(s) may have an impact on the plant development, growth and health as well as make it easier for the pathogen to infect the plant. Although these mutations might not be fatal, they can nevertheless have pleiotropic consequences such as dwarfism, food shortage and the inhibition of genes involved in replication (Ahmad *et al.* 2020; Pavan *et al.* 2010). To overcome these difficulties, new *S* variants (Bastet *et al.* 2018) must be created and inserted into the plants, or base modification must be used (Rodríguez-Leal *et al.* 2017). However, multiple investigations revealed that plant development was unaffected even after indels for mutation in the translational initiation factor (Macovei *et al.* 2018; Pyott *et al.* 2016). However, it still needs more experimental proof to be supported.

Another potential drawback of the CRISPR-Cas9 technology that has drawn attention from scientists is off-target mutation. A point mutation, insertion, or translocation that happened in the genome at undesirable site after GE is referred to as an off-target mutation (Liang *et al.* 2018; Liang *et al.* 2017). Depending on the type of mutation, these undesirable changes can affect how competent the system can change the structure and function of genes, or even influence the behavior of cells. Initially, the scientists were not interested in the off-target mutation in the plant system, though, because backcrosses may potentially correct or eliminate the off-target mutation (Liang *et al.* 2017). Common limitations of CRISPR-Cas system have been provided in Figure 4.2. Though, concerns about it arise from reverse genetic research as well as the production of some undesirable results, such as super weeds, etc. Recently, researchers have attempted to solve the off-target problems by either re-engineering the CRISPR components, such as Cas proteins and gRNA, or by employing computational methods.

CONCLUSION AND FUTURE PROSPECTS

The unique way to generate targeted mutations and genetic diversity in plants via GETs has brought incredible advances in both crop improvement and genetics. CRISPR-Cas9 system has emerged as a powerful tool for crop improvement due to its robustness, simplicity and versatility. Currently, due to its multiple genome-editing capabilities, e.g. gene knockout/in, insertions, deletions, replacement, fine-adjustment and activation of gene expression and other changes at any position in plants' genome, the CRISPR-Cas system has proven to be a helping hand for researchers in crop domestication and development of ideal plants with improved traits (i.e. high yield, improved quality and abiotic/biotic stresses tolerance). Despite of tremendous applications in plant biology, this system has great potential that needs to be tapped. Target prediction is most severely constrained by a lack of genetic data sets large

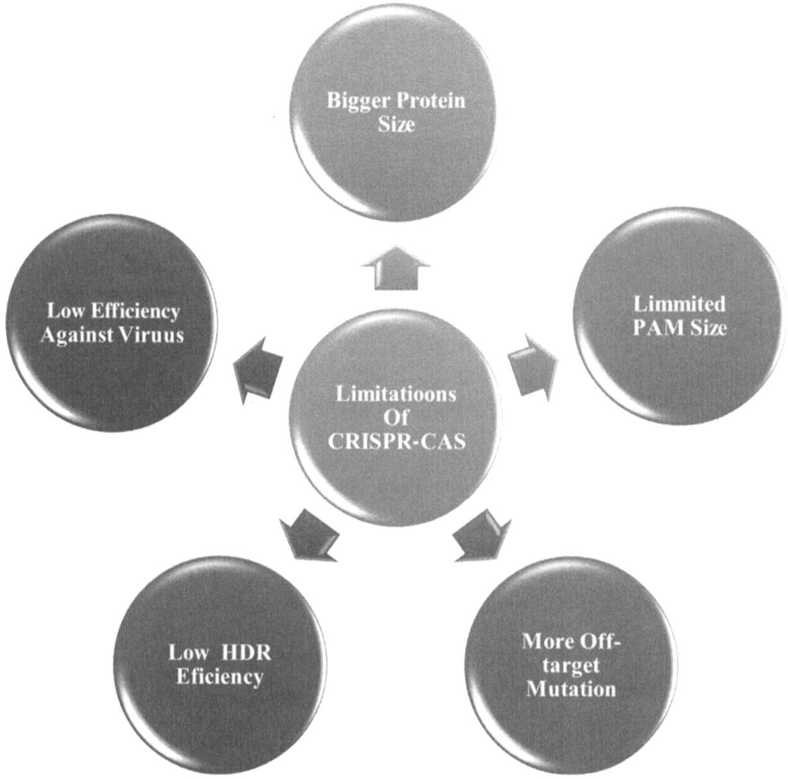

FIGURE 4.2 Limitations of CRISPR-Cas9.

enough to address the sequence specificities for various genome-editing methods. Consequently, further work is required to solve this issue. However, a little amount of target prediction is possible only by looking for a DNA-binding domain. To modulate gene expression post-transcriptionally, the CRISPR-Cas9 system can be very helpful. ABE-mediated genome editing will be particularly helpful for producing precise point mutations and deletions with fewer indels. Future DSB-mediated genome-editing techniques may make use of synthetic plant genomes and chromosome engineering. These genetic switches can be built into the genetic circuit to turn on and off a certain characteristic or set of genetic processes. It could be particularly interesting to introduce a 'genome-editing' library, including experimental references and in-silico prediction data of model organisms. A researcher might be able to identify appropriate genome-editing tools for a difficult gene in the database.

REFERENCES

Abudayyeh, O. O., Gootenberg, J. S., Konermann, S., Joung, J., Slaymaker, I. M., Cox, D. B., ... Minakhin, L. (2016). C2c2 is a single-component programmable RNA-guided RNA-targeting CRISPR effector. *Science, 353*(6299), aaf5573.

Ahmad, S., Cheema, H. M. N., Khan, A. A., Khan, R. S. A., & Ahmad, J. N. (2019). Resistance status of Helicoverpa armigera against Bt cotton in Pakistan. *Transgenic Research, 28*(2), 199–212.

Ahmad, S., Wei, X., Sheng, Z., Hu, P., & Tang, S. (2020). CRISPR/Cas9 for development of disease resistance in plants: recent progress, limitations and future prospects. *Briefings in Functional Genomics, 19*(1), 26–39.

Ali, Z., Abul-Faraj, A., Li, L., Ghosh, N., Piatek, M., Mahjoub, A., ... Voytas, D. F. (2015). Efficient virus-mediated genome editing in plants using the CRISPR/Cas9 system. *Molecular Plant, 8*(8), 1288–1291.

Alwin, S., Gere, M. B., Guhl, E., Effertz, K., Barbas III, C. F., Segal, D. J., ... Cathomen, T. (2005). Custom zinc-finger nucleases for use in human cells. *Molecular Therapy, 12*(4), 610–617.

Bastet, A., Lederer, B., Giovinazzo, N., Arnoux, X., German-Retana, S., Reinbold, C., ... Gersch, S. (2018). Trans-species synthetic gene design allows resistance pyramiding and broad-spectrum engineering of virus resistance in plants. *Plant Biotechnology Journal, 16*(9), 1569–1581.

Bastet, A., Robaglia, C., & Gallois, J.-L. (2017). eIF4E resistance: natural variation should guide gene editing. *Trends in Plant Science, 22*(5), 411–419.

Bitinaite, J., Wah, D. A., Aggarwal, A. K., & Schildkraut, I. (1998). Fok I dimerization is required for DNA cleavage. *Proceedings of the National Academy of Sciences, 95*(18), 10570–10575.

Boch, J., & Bonas, U. (2010). Xanthomonas AvrBs3 family-type III effectors: Discovery and function. *Annual Review of Phytopathology, 48*, 419–436.

Bonas, U., Stall, R. E., & Staskawicz, B. (1989). Genetic and structural characterization of the avirulence gene avrBs3 from Xanthomonas campestris pv. vesicatoria. *Molecular and General Genetics MGG, 218*(1), 127–136.

Carlson, D. F., Tan, W., Lillico, S. G., Stverakova, D., Proudfoot, C., Christian, M., ... Fahrenkrug, S. C. (2012). Efficient TALEN-mediated gene knockout in livestock. *Proceedings of the National Academy of Sciences, 109*(43), 17382–17387.

Carroll, D. (2011). Genome engineering with zinc-finger nucleases. *Genetics, 188*(4), 773–782.

Cong, L., Ran, F. A., Cox, D., Lin, S., Barretto, R., Habib, N., . . . Marraffini, L. A. (2013). Multiplex genome engineering using CRISPR/Cas systems. *Science, 339*(6121), 819–823.

Damalas, C. A., & Eleftherohorinos, I. G. (2011). Pesticide exposure, safety issues, and risk assessment indicators. *International Journal of Environmental Research and Public Health, 8*(5), 1402–1419.

Ding, D., Chen, K., Chen, Y., Li, H., & Xie, K. (2018). Engineering introns to express RNA guides for Cas9-and Cpf1-mediated multiplex genome editing. *Molecular Plant, 11*(4), 542–552.

Gao, C. (2018). The future of CRISPR technologies in agriculture. *Nature Reviews Molecular Cell Biology, 19*(5), 275–276.

Gasiunas, G., Barrangou, R., Horvath, P., & Siksnys, V. (2012). Cas9–crRNA ribonucleoprotein complex mediates specific DNA cleavage for adaptive immunity in bacteria. *Proceedings of the National Academy of Sciences, 109*(39), E2579–E2586.

Gupta, A., Christensen, R. G., Rayla, A. L., Lakshmanan, A., Stormo, G. D., & Wolfe, S. A. (2012). An optimized two-finger archive for ZFN-mediated gene targeting. *Nature Methods, 9*(6), 588–590.

Joung, J. K., & Sander, J. D. (2013). TALENs: a widely applicable technology for targeted genome editing. *Nature Reviews Molecular Cell Biology, 14*(1), 49–55.

Kaur, N., Alok, A., Kaur, N., Pandey, P., Awasthi, P., & Tiwari, S. (2018). CRISPR/Cas9-mediated efficient editing in phytoene desaturase (PDS) demonstrates precise manipulation in banana cv. Rasthali genome. *Functional & Integrative Genomics, 18*, 89–99.

Kim, E., Koo, T., Park, S. W., Kim, D., Kim, K., Cho, H.-Y., ... Kim, S. (2017). In vivo genome editing with a small Cas9 orthologue derived from Campylobacter jejuni. *Nature Communications, 8*(1), 14500.

Liang, Y., Han, Y., Wang, C., Jiang, C., & Xu, J.-R. (2018). Targeted deletion of the USTA and UvSLT2 genes efficiently in Ustilaginoidea virens with the CRISPR-Cas9 system. *Frontiers in Plant Science, 9*, 699.

Liang, Z., Chen, K., Li, T., Zhang, Y., Wang, Y., Zhao, Q., ... Ran, Y. (2017). Efficient DNA-free genome editing of bread wheat using CRISPR/Cas9 ribonucleoprotein complexes. *Nature Communications, 8*(1), 1–5.

Liang, Z., Chen, K., Zhang, Y., Liu, J., Yin, K., Qiu, J.-L., & Gao, C. (2018). Genome editing of bread wheat using biolistic delivery of CRISPR/Cas9 in vitro transcripts or ribonucleoproteins. *Nature Protocols, 13*(3), 413–430.

Liu, L., Chen, P., Wang, M., Li, X., Wang, J., Yin, M., & Wang, Y. (2017). C2c1-sgRNA complex structure reveals RNA-guided DNA cleavage mechanism. *Molecular Cell, 65*(2), 310–322.

Liu, L., Li, X., Wang, J., Wang, M., Chen, P., Yin, M., ... Wang, Y. (2017). Two distant catalytic sites are responsible for C2c2 RNase activities. *Cell, 168*(1–2), 121–134. e112.

Liu, Q., Segal, D. J., Ghiara, J. B., & Barbas III, C. F. (1997). Design of polydactyl zinc-finger proteins for unique addressing within complex genomes. *Proceedings of the National Academy of Sciences, 94*(11), 5525–5530.

Liu, Y., Ma, S., Wang, X., Chang, J., Gao, J., Shi, R., ... Zhao, P. (2014). Highly efficient multiplex targeted mutagenesis and genomic structure variation in Bombyx mori cells using CRISPR/Cas9. *Insect Biochemistry and Molecular Biology, 49*, 35–42.

Lowder, L. G., Zhang, D., Baltes, N. J., Paul III, J. W., Tang, X., Zheng, X., ... & Qi, Y. (2015). A CRISPR/Cas9 toolbox for multiplexed plant genome editing and transcriptional regulation. *Plant Physiology*, *169*(2), 971–985.

Lowder, L., Malzahn, A., & Qi, Y. (2016). Rapid evolution of manifold CRISPR systems for plant genome editing. *Frontiers in Plant Science*, *7*, 1683.

Macovei, A., Sevilla, N. R., Cantos, C., Jonson, G. B., Slamet-Loedin, I., Čermák, T., ... Chadha-Mohanty, P. (2018). Novel alleles of rice eIF4G generated by CRISPR/Cas9-targeted mutagenesis confer resistance to Rice tungro spherical virus. *Plant Biotechnology Journal, 16*(11), 1918–1927.

Mali, P., Yang, L., Esvelt, K. M., Aach, J., Guell, M., DiCarlo, J. E., ... Church, G. M. (2013). RNA-guided human genome engineering via Cas9. *Science, 339*(6121), 823–826.

Mani, M., Smith, J., Kandavelou, K., Berg, J. M., & Chandrasegaran, S. (2005). Binding of two zinc finger nuclease monomers to two specific sites is required for effective double-strand DNA cleavage. *Biochemical and Biophysical Research Communications, 334*(4), 1191–1197.

Mohanta, T. K., Bashir, T., Hashem, A., Abd_Allah, E. F., & Bae, H. (2017). Genome editing tools in plants. *Genes, 8*(12), 399.

Mushtaq, M., Ahmad Dar, A., Skalicky, M., Tyagi, A., Bhagat, N., Basu, U., . . . Dar, T.-U.-H. (2021). CRISPR-based genome editing tools: Insights into technological breakthroughs and future challenges. *Genes, 12*(6), 797.

Pabo, C. O., Peisach, E., & Grant, R. A. (2001). Design and selection of novel Cys2His2 zinc finger proteins. *Annual Review of Biochemistry, 70*(1), 313–340.

Pâques, F., & Duchateau, P. (2007). Meganucleases and DNA double-strand break-induced recombination: perspectives for gene therapy. *Current Gene Therapy, 7*(1), 49–66.

Pavan, S., Jacobsen, E., Visser, R. G., & Bai, Y. (2010). Loss of susceptibility as a novel breeding strategy for durable and broad-spectrum resistance. *Molecular Breeding, 25*(1), 1–12.

Pavletich, N. P., & Pabo, C. O. (1991). Zinc finger-DNA recognition: crystal structure of a Zif268-DNA complex at 2.1 Å. *Science, 252*(5007), 809–817.

Pests, F. K. P. (2015). *Diseases at Bay: Experts Focus on Global Measures*. http://www.fao.org/news/story/en/item/280489/icode/ (accessed on 15 February 2022)
Pyott, D. E., Sheehan, E., & Molnar, A. (2016). Engineering of CRISPR/Cas9-mediated potyvirus resistance in transgene-free Arabidopsis plants. *Molecular Plant Pathology, 17*(8), 1276–1288.
Ran, F. A., Cong, L., Yan, W. X., Scott, D. A., Gootenberg, J. S., Kriz, A. J., ... Makarova, K. S. (2015). In vivo genome editing using Staphylococcus aureus Cas9. *Nature, 520*(7546), 186–191.
Rodríguez-Leal, D., Lemmon, Z. H., Man, J., Bartlett, M. E., & Lippman, Z. B. (2017). Engineering quantitative trait variation for crop improvement by genome editing. *Cell, 171*(2), 470–480. e478.
Sander, J. D., & Joung, J. K. (2014). CRISPR-Cas systems for editing, regulating and targeting genomes. *Nature Biotechnology, 32*(4), 347–355.
Sandhya, D., Jogam, P., Allini, V. R., Abbagani, S., & Alok, A. (2020). The present and potential future methods for delivering CRISPR/Cas9 components in plants. *Journal of Genetic Engineering and Biotechnology, 18*, 1–11.
Smargon, A. A., Cox, D. B., Pyzocha, N. K., Zheng, K., Slaymaker, I. M., Gootenberg, J. S., . . . Makarova, K. S. (2017). Cas13b is a type VI-B CRISPR-associated RNA-guided RNase differentially regulated by accessory proteins Csx27 and Csx28. *Molecular Cell, 65*(4), 618–630. e617.
Smith, J., Grizot, S., Arnould, S., Duclert, A., Epinat, J.-C., Chames, P., ... Bravo, J. (2006). A combinatorial approach to create artificial homing endonucleases cleaving chosen sequences. *Nucleic Acids Research, 34*(22), e149–e149.
Thakore, P. I., & Gersbach, C. A. (2015). Genome engineering for therapeutic applications. In *Translating Gene Therapy to the Clinic* (pp. 27–43): Elsevier.
Tek, M. I., & Budak, K. (2022). A New Approach to Develop Resistant Cultivars Against the Plant Pathogens: CRISPR Drives. *Frontiers in Plant Science, 13*. e889497
Voytas, D. F. (2013). Plant genome engineering with sequence-specific nucleases. *Annual Review of Plant Biology, 64*, 327–350.
Wah, D. A., Bitinaite, J., Schildkraut, I., & Aggarwal, A. K. (1998). Structure of Fok I has implications for DNA cleavage. *Proceedings of the National Academy of Sciences, 95*(18), 10564–10569.
Wilde, J. J., Aida, T., Del Rosario, R. C., Kaiser, T., Qi, P., Wienisch, M., ... Feng, G. (2021). Efficient embryonic homozygous gene conversion via RAD51-enhanced interhomolog repair. *Cell, 184*(12), 3267–3280. e3218.
Yin, K., Han, T., Liu, G., Chen, T., Wang, Y., Yu, A. Y. L., & Liu, Y. (2015). A geminivirus-based guide RNA delivery system for CRISPR/Cas9 mediated plant genome editing. *Scientific Reports, 5*(1), 1–10.
Zaidi, S. S.-e.-A., Vanderschuren, H., Qaim, M., Mahfouz, M. M., Kohli, A., Mansoor, S., & Tester, M. (2019). New plant breeding technologies for food security. *Science, 363*(6434), 1390–1391.
Zhang, Y., Liang, Z., Zong, Y., Wang, Y., Liu, J., Chen, K., ... Gao, C. (2016). Efficient and transgene-free genome editing in wheat through transient expression of CRISPR/Cas9 DNA or RNA. *Nature Communications, 7*(1), 1–8.

5 Developments in Plant Viral Resistance Using CRISPR-Cas System

INTRODUCTION

Numerous plant diseases, which may be induced by a broad variety of plant pathogens such as bacteria, fungi, nematodes and viruses, are the root problem of massive economic devastation around the globe (Prasher & Sharma, 2022). About half of all plant diseases are caused by viruses and it is thought that viral infections of crop varieties cost the world more than US$30 billion each year. So far, according to the study of viral genome analysis, over 1,500 different species of plant viruses have been recognized and classified into 26 different families. Plant viruses belong to a category of obligate intracellular parasites that have low capabilities for genetic coding and are highly dependent on their hosts for the successful completion of their life cycle (Boro *et al.* 2022).

Plant viruses are nucleoprotein complexes and are dependent on their host cells for most of their reproduction. Various novel plant diseases are caused by viruses. This is mostly due to the flexibility of viruses to adapt the changing environmental conditions and efficient dispersion of viruses that is enabled through vector transmission (Karmakar *et al.* 2022). Most economically important plant species are susceptible to infection by viruses, which may result in devastating viral diseases. These diseases are directly responsible for considerable drop in both the quantity and quality of harvests across the globe. It is believed that plant diseases account for a loss of fifteen percent of the world's total agricultural yield, with viruses being responsible for thirty percent of that total. So, plant viruses threaten the safety of the food supply worldwide and the agriculture production required to feed the world's population, which is growing all the time (Javed *et al.* 2022).

The life sciences are undergoing a paradigm shift as a direct result of recent developments in genome editing methods including sequence-specific nucleases (SSNs). This change is mostly caused by clustered regularly interspaced palindromic repeats-associated nuclease systems (CRISPR-Cas). In prokaryotic organisms, the innate immune mechanism known as CRISPR-Cas is responsible for conferring resistance to the nucleic acids of other organisms (Malik *et al.* 2022). As compared to other SSNs, such as zinc finger nucleases (ZFNs) and transcription activator-like effector nucleases (TALENs), the CRISPR-Cas method of editing the genome makes it much simpler to engineer specific targetable nuclease systems. This is one of the

DOI: 10.1201/b22901-5

many advantages of using the CRISPR-Cas approach. Editing the genome using CRISPR-Cas provides fresh alternatives to traditional plant breeding and transgenic techniques of agricultural enhancement, which may help mitigate the ever-present threat posed by the Malthusian check (Fenibo *et al.* 2022).

The CRISPR and CRISPR-associated (Cas) systems play a crucial role in the adaptive immune response to invading nucleic acids and have their roots in bacteria and archaea. Recent advances in our understanding of the CRISPR-Cas systems have enabled the development of new tools for editing both endogenous and exogenous portions of DNA or RNA in a wide range of organisms. To this day, all that is necessary for a simplified CRISPR system is a guide RNA (gRNA) that can be quickly synthesized with a Cas effector protein (Pires *et al.* 2022). Based on the sequence, structure and function of Cas proteins, two distinct types of CRISPR-Cas systems may be separated from one another. All four forms of CRISPR-Cas systems rely on a similar multiprotein effector complex; however, type I systems employ a different complex than types III and IV. CRISPR-Cas system is classified as type I to type VI. Type I systems are characterized based on the occurrence of signature protein Cas3, a protein which contains both DNase and helicase domains used to degrade the target. Type II CRISPR/Cas systems utilize Casl, Cas2, Cas9 and a fourth protein (Csn2 or Cas4), whereas the type III CRISPR/Cas systems comprise the Casl0 with an indistinct role. The type II CRISPR/Cas system originates from *S. pyogenes* and comprises three components: the CRISPR RNA (crRNA), trans-activating crRNA (tracrRNA), and a Cas9 protein (Kulabhusan *et al.* 2022).

As a result, class 2 CRISPR-Cas systems have gained widespread acceptance and are now being applied for the alteration and detection of nucleic acids. The type II and type V Cas proteins are used to edit DNA, whereas type VI Cas proteins are used to edit RNA (Kerber-Diaz *et al.* 2022). CRISPR-Cas-mediated genome editing has widely been adapted over the course of the last several years. The ultimate goal of this robust technology is to develop resistance against viruses and other diseases. In order to prevent the replication of invading viruses and the spread of infection, one strategy involves focusing on destroying and interfering the viral genome. The second strategy involves modifying host susceptibility factors, which are necessary for the infection and life cycle of the virus, in order to boost plant immunity and prevent virus invasion (Hillary & Ceasar, 2022).

MECHANISMS OF VIRAL INFECTION IN PLANTS

Plants, whether they are agricultural, medicinal, or decorative, are susceptible to infection by viruses. It is not necessary for the virus to infect more than one cell before it may spread. However, since viruses are incapable of acting on their own, they must use the machinery of the cell that they have infected to generate copies of themselves (Yue *et al.* 2022). At some point, offspring viruses will be disseminated to neighboring cells and the cycle will begin again from the beginning. The virus will soon be able to enter the plant's vascular system and spread far beyond the site of initial infection, infecting not just the plant's roots but also its young leaves. The immune systems of both humans and animals are well-suited to combat viral infections. Silencing genes may be a powerful tool in the struggle against viruses in both plants and insects (Yang & Liu, 2022). This process considers a virus to be analogous to a gene whose expression has become unregulated. Therefore, to switch gene off, plant cells chop up the viral RNA into very little bits. This puts an end to the infection by inhibiting the capacity of the virus's RNA to replicate further or by preventing the localized spread. One fascinating feature of this system is that infected cells may send a signal to neighboring cells, which then activates antiviral defense before the virus has even reached its target. Normally, this method relies on a collection of proteins that function together to control genes regulation inside the cells (Yang *et al.* 2022).

Plants are at a disadvantage, since viruses do not give up lightly and have a few more tricks up their sleeves. In order to combat the process of gene silencing that is carried out by plants, viruses are able to develop specialized proteins that have the ability to interfere with plant defense mechanisms. This may be accomplished via a variety of methods (Shukla *et al.* 2022). It is always one or more of the components that are engaged in the defensive response of the plant that will serve as the target for destruction or inactivation. Who comes out on top in this conflict will primarily be determined by the equilibrium that exists between the plant's ability to silence antiviral genes and the virus's ability to block this process (Wu *et al.* 2022). During the last decade, researchers have been putting a lot of effort into understanding the factors that will determine the result of this struggle. Together with his team, Dr. Hernan Garcia-Ruiz is making significant strides in research that will help in the discovery and characterization of cellular elements that condition a plant's sensitivity to viral infections. As part of their research endeavor, they are investigating the processes of antiviral gene silencing by using model plants and RNA viruses. This research was found to elucidate the processes of antiviral gene silencing by clarifying how viral RNAs are distinguished from cellular (untargeted) RNA (Lopez-Gomollon & Baulcombe, 2022).

HOW CAN PLANTS SUSTAIN THEMSELVES AGAINST VIRUSES?

Gene silencing is known to be used by plants as a defense system toward infection, but the signal that triggers this response in the presence of a virus has yet to be identified. Furthermore, it has been studied that once a plant identifies a virus, its priority is to sever and destroy the viral RNA to prohibit the infection from spreading to other plants (Jones, 2022). This complicated process is dependent, in a significant way, on a sequence of proteins that can specifically target and destroy viral RNA. It was previously believed that the method that plants use to defend themselves against viral infections is simple. However, it is not as simple as figuring out which proteins are involved in the process. The reactions taking place in plants are complex one as plants could have *de novo* pathway other than the traditional one as a backup plan (Akbar *et al.* 2022).

When such a situation arises, the second set of proteins can safeguard the plant even if sometimes it is not as effective as the main system. In contrast, it seems that various proteins are more efficient than others in protecting specific plant parts, such as the leaves or the flowers, while other proteins have a powerful response regardless of whatever plant component is being targeted (Popoola *et al.* 2022). While others are more effective when working in conjunction with a few other proteins, others may function on their own. In the end, whether the fight against the virus is effective, it relies on a delicate balancing act of all of these processes. Understanding this mechanism is crucial for learning how plants naturally defend themselves against viruses, why certain viruses are more virulent in specific hosts and how sustainable antiviral resistance approaches may be used in agriculture (Mangukia *et al.* 2022).

RETALIATION AGAINST VIRUSES

The plant's defense might provoke a reaction from the virus, which the virus will then utilize to launch its own counterattack (Balan & Kang, 2022). This is accomplished by a combination of different techniques, one of which is the hunting down and destroying of all plant proteins that are engaged in their defense system. Nevertheless, every virus has its own unique set of destructive capabilities, and the specific processes that underlie each infection which have not yet been fully understood (Banerjee *et al.* 2022).

THE COMPLEXITY OF VIRAL INFECTIONS IN PLANTS

On rare occasions, different virus variants may infect a plant simultaneously. In the 1970s, scientists in Kansas and Nebraska discovered a fatal necrosis in maize by maize chlorotic mottle virus (MCMV) that interacts synergistically with sugarcane mosaic virus, wheat stripe mosaic virus and Johnson grass mosaic virus, causing this disease. Disease indicators include cobs that are deformed or decaying, smaller, perhaps sterile plants and yellow leaves that have dried at the edges. However, Dr. Garcia-Ruiz and his team discovered plants in the fields of Kenyan farmers in 2017 that had bright yellow stripes with green edges. When compared to other symptoms of maize lethal necrosis, these stripes stood out (Patra *et al.* 2022). Two, three, or perhaps four different viruses, including the maize yellow mosaic virus, were found in the plants in Kenya.

The fact that sorghum, Napier grass and even other kinds of plants may also become infected with this disease makes it exceedingly difficult to stop its spread. Understanding how the viruses work together to combat the plants' reaction is an important feature that requires more investigation. The work that Dr. Garcia-Ruiz and his colleagues have done is just the beginning, since, from a pragmatic perspective, the disease is spreading continuously and has become a big problem for farmers (Van Rythoven, 2022). To understand the role of silencing in maize lethal necrosis, further study is needed. This will allow for more precise viral diagnostics and a new assessment method (Ghattargi *et al.* 2022).

MECHANISMS OF VIRUS RESISTANCE IN PLANTS

Plants have both active and passive defense systems, the former one is dependent on the preexisting barriers such as the hard cell wall. Active defense systems are triggered upon the identification of viruses or other pathogens. The hypersensitive response (HR) is the response that is the most linked with an effective defense

(Sharma *et al.* 2022). During this process, cells that are located in close proximity to the foremost infected area may undergo quickly-caused cell death program. The virus must first attach specifically to the plant, and this pairing often occurs by way of the dominant gene products of the plant's resistance genes (*R* genes) (Rai *et al.* 2022). Specific identification of the virus is required to initiate this reaction. Only a few numbers of single dominant *R* resistance genes protecting plants against various pathogens have been cloned and sequenced so far. Several plant species have been used to isolate these genes. Interestingly, despite the diversity of pathogen taxa (which may include viruses, bacteria, fungus, nematode and insect pathogens), proteins with a nucleotide-binding site and a leucine-rich repeat (NBLRR) are what the largest cluster of *R* genes codes for. Until yet, all of the *R* genes that provide protection against viruses have been found in this group (Akbar *et al.* 2022; Khan *et al.* 2022). In addition to their roles in protein–protein, protein–peptide ligand and carbohydrate–protein binding, LRR domains have been shown to have a function in protein–protein binding. While it's tempting to assume that *R* gene products have some kind of interaction with other components (whether host- or virus-expressed), this has yet to be shown. This also holds true for the NB domain, suggesting that *R* gene products may bind GTP or ATP, but the significance of this is still being investigated. The large group of NB-LRR resistance genes may be further subdivided on the basis of the N-terminal domain (Abdelkhalek *et al.* 2022). This might be a leucine zipper (LZ-NB-LRR), a coiled-coil (CC) domain (CC-NB-LRR) or a TIR domain (with homology to the intracellular signaling domains of the *Toll* gene in Drosophila and the human interleukin (IL)-1 receptors) (TIR-NB-LRR) (Saikia *et al.* 2022).

It is plainly evident that functional research is necessary to learn how these *R* gene products operate and trigger an effective response to invading viruses. A comparison of the ever-increasing number of *R* genes has shown that they originated in long-lost gene families and progressed via duplication, mutation and recombination (Ishikawa *et al.* 2022). This is due to the fact that all *R* genes are connected to one another, independent of the specific plant pathogen they are designed to combat. *R* genes are usually found as part of a larger gene cluster that also comprises their homologs. The related homologs' function has, for the most part, been unknown until now. The availability of cloned *R* genes will allow for not only a deeper understanding of the mechanisms behind plant defenses but also their application to plants of other species. The *Sw-5* gene in tomatoes, which confers resistance to several *topoviruses*, is one example of many *R* genes that have been proven to be functional in a range of host plants (Saberi *et al.* 2022).

MECHANISMS

For viruses to complete their lifespans, they must accomplish a variety of things, like access to plant cells, unwrap their nucleic acid, translate their viral genome, copy one's viral nucleic acid, put together their progeny virions, move from cell to cell, move through the plant's system and move from one plant to another. Most of the time, plant viruses cause an infection by splitting through the plant cells and attempting to enter a living cell through injuries made by mechanical damage or vectors like insects and nematodes (LaTourrette & Garcia-Ruiz, 2022). In comparison to animal viruses, it is unknown how plant viruses specifically infiltrate plant cells; there isn't one technique that plant viruses exploit. This is because the evolution of viruses in plants occurred separately from that of viruses in mammals. When virions enter a plant cell that is vulnerable to infection, the genome is generally released from the capsid, which would be in the cytoplasm of a plant. This process takes place when the virus enters the plant cell. According to the findings of the current study, the uncoating mechanism does not seem to be host dependent. For instance, TMV and Tobacco yellow mottle virus were shown to be uncoated in host plants as well as nonhost plants. Even though this discovery has not been verified just yet, it provides support for the idea (Kaur *et al.* 2022). Initial viral components, including such viral copies as well as other virus-specific proteins, may be transcribed from the mRNAs that code for utilizing the genome once it is made available (Severin, 2022).

After this point, the virus is confronted with a variety of limitations imposed by the host, and it also needs the participation of many host proteins, which are generally for use in the infection cycle of the virus (Cherusseri *et al.* 2022). As a result, a virus's ability to infect a plant depends on a series of connections that are suitable with a subset of viral gene products. The absence of a necessary host ingredient or a variation leading to mismatch has traditionally been postulated as the cause of recessively inherited disease-resistant plants.. On the other hand, research into a variety of plant photosystems has uncovered that dominant resistance develops after an explicit recognition event among host and viral elements, which then triggers host disease resistance (Payne, 2022). The biochemistry behind this recognition event is still not completely understood, despite the availability of genetic systems that have been well defined and the extensive research that has been conducted in this field. The genes that contribute to this reaction are likely to be dominant or incompletely dominant, unless the resistant response is caused by the depression of a defensive system. In this case, the genes that contribute to this response are likely to be recessive. Although the majority of known viral resistance mechanisms seem to target virus multiplication or movement, passive or active resistance may theoretically work at any step of the life cycle of the virus (Di Mattia *et al.* 2022).

It is currently difficult to precisely assess levels of viral uptake in asynchronous infections of intact tissue due to the technological challenges involved (as opposed to protoplasts). Even with the use of fluorescent reporter genes, it is difficult to determine the amount to viral increase due to replication and translation, as result of fluctuations in virus mobility (Qin *et al.* 2022). Multiple pieces of evidence point to the possibility that the amount of viral accumulation may influence the virus's capacity to spread throughout the body. For instance, the systemic mobility of the Barley stripe mosaic virus may be determined by the quantity of the protein that is made by RNA 3 of the virus. Additionally, dose dependence has been shown in a variety of viral–host interactions, such as when a virus infects a bacterium. So, before jumping to the conclusion, that the molecular flaw that results in resistance directly impacts the stage of the viral infection cycle at which the defect, i.e. resistance, is found. (Fan *et al.* 2023).

BREEDING FOR RESILIENCE TO DISEASE AND ANTIVIRUS PROTECTION

Durable disease resistance is defined as "resistance that has remained effective while a cultivar possessing it has been widely cultivated in an environment favoring the disease" (Gupta *et al.* 2022). When it comes to naturally existing *R* genes, this word is the most helpful for doing retrospective analyses when applied to them. Our current knowledge of the underlying biology does not yet permit us to make any accurate estimate. Despite this, there are significant discrepancies across the *R* genes in terms of their durability when used in agriculture and this is the case even if an effort is made to correct for changes in the intensity of the use of the genes (Wani *et al.* 2022). Despite of the theory that monogenic dominant resistance is inevitably fragile, there are *R* genes that continue to be beneficial for many decades or even longer. A dominant *I* gene, which gives immunity to BCMV and numerous other diseases, has been used in snap bean breeding operations since the 1930s (Hoyos-Villegas *et al.* 2022).

This *R* gene is found in *Phaseolus vulgaris* and even though there have been reports of isolates that produce systemic necrosis, none of the pathotypes of the viruses regulated by the genes has been able to get around the resistance and cause mosaic disease. Even though the necrotic response has the potential to be more devastating than the mosaic response, the gene is nevertheless extensively used, since it prevents the viruses from being transmitted via the seed (Rubiales *et al.* 2022). There has been a minimum of 50 years of examples of widespread use of recessive *R* genes in both monocots and dicots. Recessive sensitivity may theoretically remain longer than dominant resistance.

Thus, no definitive tests of this hypothesis have been conducted yet. In addition, it was suggested that resilience or tolerances governed by a significant number of genes, each with a different but potentially overlapping impact (horizontal resilience), would be more robust than resistance governed by genes that have a greater localized influence on the organisms (vertical resistance) (Wulff & Krattinger, 2022). There is currently no conclusive response to the question of whether genes that differ with respect to either horizontal or vertical resistivity really constitute independent processes (Anuragi *et al.* 2022).

If resistance is the consequence of a cumulative effect caused by many overlapping *R* genes or mechanisms, then it will take multiple modifications for a virus to overcome this effect and spread. Very few studies have examined polygenic resistance or tolerance reactions to plant viruses, as has been previously mentioned (Klymiuk *et al.* 2022). In addition, there were even fewer examples of this resilience being passed on to other varieties of crops. Homozygotes for two recessive *R* genes, *sbm*1 and *sbm*2, have been found in peas to exhibit particularly long-lasting resistance to PSbMV.

The mechanisms by which resistance persists remain a mystery, even though the *Tm22* gene's survival may be due to its reduced efficiency as a variation inside an avirulence gene. *Rx* and *Ry* in potatoes and *pvr*1 in pepper are illustrations of viral *R* genes that function in cells (Ishikawa *et al.* 2022). Having those genes seems to provide very long-lasting protection against the virus. It is more likely that pathogenicity will not emerge in a viral population if avirulent viruses have little to no chance to proliferate in infected tissues. Viruses benefit from maintaining pathogenicity, since it increases their chances of survival. Different factors, including the demography of the viral populace, the types and rates of changes required for pathogenicity and the distribution of virulent isolates, contribute to the virus's persistence. In the context of *R* genes, this is true with both dominant and recessive forms (Rahman *et al.* 2022). Various methods of developing disease resistance are given in Table 5.1.

USING CRISPR FOR VIRAL DISEASE RESISTANCE IN PLANTS

The life sciences are undergoing a paradigm shift as a direct result of recent developments in genome editing methods including sequence-specific nucleases (SSNs). In particular, clustered regularly interspaced palindromic repeats-associated nuclease systems (CRISPR-Cas) are driving this change (Karmakar *et al.* 2022). In prokaryotic organisms, the innate immune mechanism known as CRISPR-Cas is responsible for conferring resistance to the nucleic acids of other organisms. In contrast to other alternative SSNs, such as DNA–protein recognition zinc finger nucleases (ZFNs) and transcription activator-like effector nucleases (TALENs), the CRISPR-Cas method of editing the genome makes it much simpler to engineer specific targetable nuclease systems (Shahriar *et al.* 2021). This is one of the many advantages of using the CRISPR-Cas method (Tripathi *et al.* 2021). The plant scientific community has not accepted ZFNs and TALENs to a large extent, which may be owing to the cumbersome need that two distinct DNA-binding proteins, each having a C-terminal FokI nuclease module, must surround the targeted region. Editing the genome using CRISPR-Cas provides fresh alternatives to traditional techniques of agricultural enhancement, such as plant breeding and transgenic technology (Tyagi *et al.* 2021).

The CRISPR-Cas systems in prokaryotes can integrate short, foreign transcripts from nucleic acids that are invading the host genome into a CRISPR array. CRISPR-Cas systems I, II and III make use of their own unique molecular processes in order to detect target nucleic acids and commence the cleavage process (Mubarik *et al.* 2021). The protospacer adjacent motif (PAM) is a short sequence motif adjacent to the target sequence that is essential for target identification and cleavage in type I and type II systems. In type II systems, transcripts of intruding nucleic acids from the CRISPR array are later transcribed as CRISPR RNA (crRNA), which is then undergoes post transcriptional modification and processed into a complex with trans-activating RNA (tracRNA) by the widely conserved endogenous RNase III. The type II system, which was isolated from *Streptococcus pyogenes* and is more widely called as CRISPR-Cas9, is the CRISPR-Cas system that is used in genome editing the most frequently today (Varanda *et al.* 2021). This is due to the ease with which the targeted site can be modified by changing a short sequence of approximately 20 nucleotides. The development of a synthetic 'linker loop' or scaffold that merges the crRNA and

TABLE 5.1
Plant Genome Engineering for Virus Resistance

Target Virus/Gene	Promoter Used	Plant Species	Effector Protein	Strategy of Transformation	Trait Improvement	Ref.
Cabbage leaf curl virus (CaLCuV) having guide RNAspCVA and pCVB being used	CaMV35S, AtU6	Arabidopsis	SpCas9	Agrobacterium GV3101	Viral-induced in vivo genetic modification (VIGM) shown using the Geminivirus	Bhattacharjee and Hallan (2022)
RSMV, TBMV and RRSV are three viruses that may infect tobacco. Strains of the rice stripe mosaic virus	CaMV35S,OsU6, AtU6	Tobacco and *Oryza sativa*	LshCas13a			Wang *et al.* (2022)
ORF1, 2, 3 and IR of BSV	OsU6	*Musa* spp. (banana)	SpCas9	EHA105 Agrobacterium	Produced banana plants immune to the naturally occurring BSV	Ma *et al.* (2022)
CP part of Caulimovirus	AtU6	Arabidopsis	SpCas9	Agrobacterium Floral dip recA strain	Resistance against CaMV	Plchová *et al.* (2022)
SgRNAs targeting BSCTV	AtU6	*Nicotiana* and *Arabidopsis*	SpCas9	Agrobacterium EHA105	Significant improvement in typical symptoms with leaf curling and a 97% decrease in infectivity	Gudeta, Igori, Belete, Kim and Moon (2022)

trace RNA into one short guide RNA (sgRNA) has significantly improved the design of numerous sgRNA repeats and made their manufacturing much simpler. CRISPR-Cas systems make use of double-strand break (DSB) repair mechanisms that are generally preserved in order to bring about beneficial alterations at particular loci of interest (Hinge *et al.* 2021). CRISPR-Cas technologies may be developed to make use of the nonhomologous end-joining (NHEJ) route in order to trigger the error-prone repair of double-strand breaks (DSB), which often results in frameshift mutations. This mechanism is referred to as the 'NHEJ pathway'. Moreover, CRISPR-Cas systems may be built to do the homology-directed repair (HDR) of a DNA template on an exogenous donor vector, which effectively enables the introduction of exogenous DNA or nucleotide sequences. In case of plant viruses, the prevalence of insect vectors, viruses spread quickly and are difficult to control. They may spread from animals to plants, making their elimination a priority in the agricultural sector (Khan *et al.* 2021). Antiviral efforts have progressed greatly during the last several decades. This new tool may be used to combat plant viruses using either DNA or RNA. DNA-binding experiments showed that the artificial zinc finger proteins (AZP) effectively inhibited the binding of the viral replication initiator protein, replication-associated protein (Rep), to the replication origin in vitro (Karmakar *et al.* 2022).

Transgenic Arabidopsis plants expressing AZP exhibited very resistant phenotypes to viral infection, with 84% of plants showing no symptoms at all. Inhibiting DNA virus replication using GETs, especially CRISPR-Cas9, has been shown to be effective, with the resulting targeted alterations resulting in resistance to viral infections in several studies (Zhang *et al.* 2022). Banana streak virus (BSV) is a plant pathogenic badnavirus of the family *Caulimoviridae* that was recently inactivated by means of the CRISPR-Cas9 system in *Musa* spp. Under water-stress circumstances, six out of eight (75%) modified events demonstrated resistance against the targeted pathogen, as compared to nonedited control plants of the banana variety Gonja Manjaya (Voloudakis *et al.* 2022). The results showed that BSV was rendered unable to produce infectious viral particles when their sequences were modified. Additional research has shown that CRISPR-Cas9 may be used to generate plant viral resistance; examples include the Tomato yellow leaf curl virus (TYLCV), the Bean yellow dwarf virus (BeYDV), the Bean streak mosaic virus (BSCTV) and others. Tobacco and tomato plants immune to TYLCV were created by targeting coat protein (CP) and Rep loci of the genome through CRISPR-Cas (Yao *et al.* 2022).

Using CRISPR-Cas9, the CP, Rep, intergenic region and long intergenic region of BeYDV and BSCTV were disrupted in tobacco and Arabidopsis to create virus-resistant plants. Less viral DNA was accumulated in the mutant plants, and they were more resistant to the specific virus (Yadav *et al.* 2022). Tobacco has also been cured of the Cotton Leaf Curl Multan virus (CLCuMuV) using Cas9 equipped with twin gRNAs to target two crucial areas of the virus's single-stranded DNA genome. Barley plants resistant to Wheat Dwarf Virus (WDV) were successfully created using CRISPR-Cas9-mediated mutagenesis. There, a vector system was developed to generate transgenic plants with a binary construct comprising four WDV-specific sgRNAs under the control of three distinct monocotyledon-specific small nuclear RNA promoters.

Recent research revealed that CRISPR mutants of cassava did not develop enough resistance to the African Cassava mosaic virus. They also uncovered a worrisome scenario in which between 33% and 48% of altered viral genomes had developed a conserved single nucleotide mutation that provided resistance to CRISPR-Cas9 cleavage (Bánfalvi *et al.* 2022). The creation of novel viruses and the evolution of CRISPR-resistant viruses is a critical issue that needs more research and more effective CRISPR systems to combat. However, the CRISPR-Cas9 system has also shown to be a useful weapon in the fight against RNA viruses (Kasanen *et al.* 2022).

Eukaryotic translation initiation factor 4E (eIF4E), also known as cap-binding protein, has been demonstrated to increase resistance against RNA-based ipomovirus and potyvirus in two separate experiments conducted on Arabidopsis and cucumber using the CRISPR-Cas9 system (Brown *et al.* 2022). However, neither study included any transgene-containing genome editing events. Similarly, CRISPR-Cas9 has been employed to target two of the five eIF4E proteins, new cap-binding protein-1 (nCBP-1) and nCBP-2, in Cassava, resulting in resistance to Cassava brown streak virus (CBSV). Other crucial genes, including tomato dicer-like 2 (*SlD*CL2), play a role in the pathways of the plant's defense system and generate virus protection (Martínez-Alonso *et al.* 2022). To verify the importance of *DCL2* in the tomato plant's defense mechanism against RNA viruses, we employed the CRISPR-Cas9 method to knock out this gene. *DCL2* is an essential part of resistance pathways against potato virus X (PVX) and tobacco mosaic virus (TMV). When infected with PVX and TMV, *DCL2* CRISPR mutants displayed typical viral symptoms, indicating a potential involvement of the *DCL2* gene in the immune response to RNA viruses (Osborne *et al.* 2022).

New CRISPR-Cas systems, including FnCas9 and CRISPR-Cas13a, formerly known as C2c2, have been used effectively to target other viruses with single-stranded RNA (ssRNA) genomes. FnCas9 and sgRNA have been used to eliminate Cucumber mosaic virus and Tobacco mosaic virus, respectively, in tobacco and Arabidopsis, with the mutant events displaying moderate disease signs and decreased accumulation of the virus. In addition, CRISPR-Cas13a is an efficient sequence-editing technique that specifically targets ssRNA. Tobacco plants developed immunity to Turnip mosaic virus (TuMV) after their genomes were edited using CRISPR-Cas13a. All of this research adds up to the conclusion that GET-mediated targeted deletion of negative regulators and/or *S* genes is the most effective method for disease-resistant plant breeding currently available (Wiebe *et al.* 2022). A general workflow for developing disease resistance using CRISPR system is given in Figure 5.1.

CRISPR-Cas-MEDIATED RESISTANCE FOR DNA VIRUSES

Transforming plants in order to make them resistant to viruses is a common practice, and it has been well documented in literature. Transgenic tobacco plants expressing the coat protein (CP) gene from the Tobacco mosaic virus (TMV) were the first plants to show delayed disease development as a result of viral resistance engineering (Khan *et al.* 2022). Cotton leaf curl disease is a most destructive disease of cotton. In a study, the efficiency of dCas9 as DNA binding protein targeting the nona-nucleotide sequence of cotton leaf curl virus (CLCuV) to inhibit virus replication was

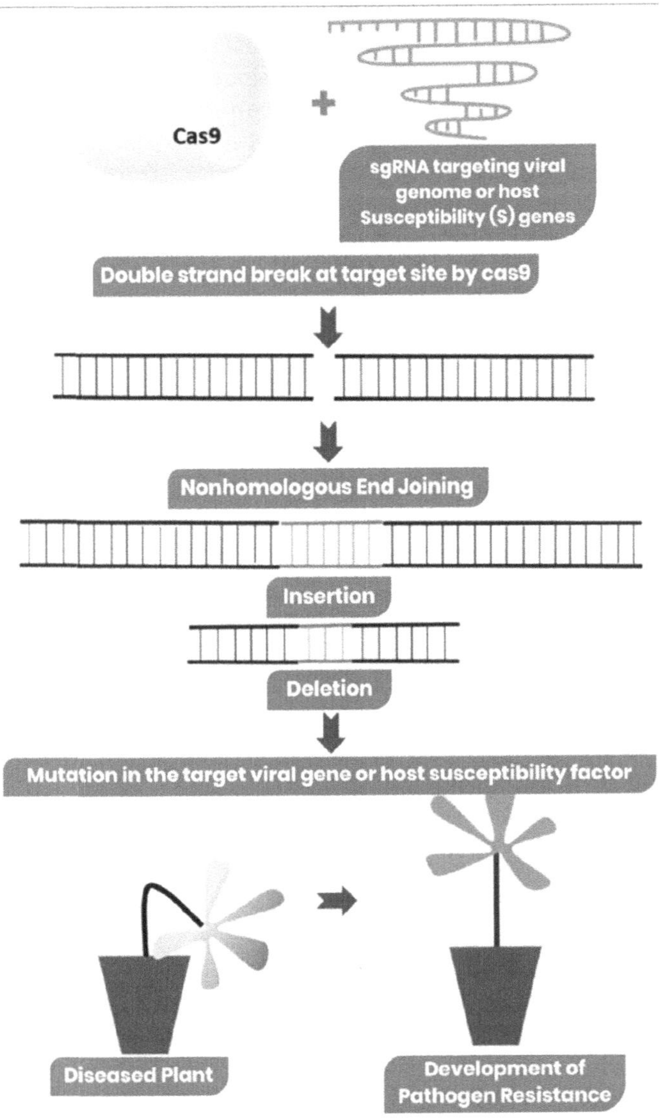

FIGURE 5.1 Workflow for developing CRISPR-mediated disease resistant plants by targeting S genes

checked. *Nicotiana benthamiana* was used to evaluate the efficiency of CRISPR/dCas9 system for viral interference. The results showed partial resistance to CLCuV and lower disease symptoms (Khan *et al.* 2019). Multiplex CRISPR/Cas mediated genome editing technique has been demonstrated and it showed that CLCuV can be controlled through this technique as the treated plants targeted the resistance of up to 60–70% (Binyamin *et al.* 2021).

Eventually, researchers at the University of Hawaii and Cornell University worked together to express the CP gene of the Papaya ringspot virus in transgenic papaya

(Carica papaya 'Rainbow') to make it resistant to the virus (PRSV). When the papaya industry in Hawaii was threatened by papaya ringspot disease, the development of the transgenic 'Rainbow' cultivar saved the day. With its capacity to target and eliminate foreign nucleic acids, a CRISPR-Cas system is a promising tool for engineering transgenic plants with resistance to viruses. *Arabidopsis thaliana* and *Nicotiana benthamiana* have been used as model organisms for the introduction of many variants of the prokaryotic CRISPR-Cas system into plants in order to impart resistance to plant diseases caused by geminiviruses (Tiwari *et al.* 2022).

The genomes of geminiviruses are circular ssDNA structures that replicate by rolling-circle mechanisms, producing dsDNA intermediates. The viral load and symptoms in transgenic *N. benthamiana* have decreased by using a CRISPR-Cas immune system designed to target the Bean yellow dwarf virus (BeYDV) (Karmakar *et al.* 2022). This was accomplished by generating mutations in the BeYDV genome and lowering copy numbers. Transgenic *N. benthamiana* infiltrated with a CRISPR-Cas immune system designed to target Beet severe curly top virus (BSCTV) attenuated the accumulation of the virus in transient experiments and induced mutations through the NHEJ repair pathway. And plants engineered to produce recombinant of the same CRISPR-Cas system in both *N. benthamiana* and *A. thaliana* showed exceptional resistance to BSCTV infection (Ghosh & Dey, 2022). These plants were able to reduce BSCTV accumulation, and the effectiveness of this suppression was associated with the amount of Cas9 expression. Transgenic *N. benthamiana* with a Tomato yellow leaf curl virus (TYLCV)-targeting CRISPR-Cas immune system dramatically decreased or eliminated symptoms. Furthermore, resistance to numerous viruses was conferred by a single sgRNA in transgenic *N. benthamiana* by targeting the highly conserved stem–loop motif of the origin of replication (ORI) (Mahas & Mahfouz, 2018).

Intragenic region (IR) of TYLCV targeting the ORI stem–loop sequence-induced resistance to TYLCV, BSCTV and Merrima mosaic virus (MeMV), all of which are geminiviruses. It seems that recombination or assortment of viral genomes causes geminiviruses to display varying pathogenicity and symptom severity. CRISPR-Cas systems developed to target the coding regions of geminiviruses have been shown to develop CRISPR-generated virus mutants with the ability to replicate and spread throughout the body (Watson *et al.*, 2018). However, interference with replication and the elimination of viral variations were seen when the noncoding IR region of geminiviruses was targeted. Numerous efforts to combat agriculturally significant ssDNA viruses using existing plant transformation technology have been fruitless. Transgenic plants produced with CRISPR-Cas systems have shown early efficacy in combating geminiviruses, suggesting that the method may be useful in combating other ssDNA viruses, such as BBTV (Gordon *et al.* 2022).

POSSIBLE CRISPR-Cas-MEDIATED RESISTANCE FOR RNA VIRUSES

Despite CRISPR-Cas-efficacy as an immune response in plants against ssDNA geminiviruses, duplication of the system targeting plant viruses with RNA genomes (most plant viruses) have yet to converge. Using a Cas endonuclease isolated from *Francisella novicida* (FnCas9), a CRISPR-Cas system was created to specifically target RNA in mammalian cells (Pandita, 2022). Guide RNAs (gRNAs) are used by the FnCas9 system to target endogenous mRNA in host cells. In Plants, tolerance to key viruses might be conferred using a plant codon-optimized variant of FnCas9 in combination with rRNA targeting ssRNA viruses (Ahmad *et al.* 2021). The class 2 type VI-A CRISPR-Cas effector is the latest RNA-targeting CRISPR-Cas system (C2c2). The C2c2 protein discovered in the bacterium *Leptotrichia shahii* may be directed to cleave matching protospacers on ssRNA targets and knock down mRNAs in bacteria. Plant virologists now have access to other CRISPR-Cas systems, including C2c2 and FnCas9, to combat plant pathogenic RNA viruses (Tuncel & Qi, 2022).

A CHALLENGE FOR THE FUTURE

The scientific community has effectively incorporated the few examples of natural genetic resistance that have been uncovered into commercial cultivars. Unfortunately, this only applies in certain circumstances. Many plant diseases, especially those caused by viruses and resulting in lower agricultural output, cannot be fought off by the plant itself. The good news is that it is possible to develop viral resistance in plants using contemporary genome editing and artificial gene silencing approaches, which in turn improves their immune systems. It's like the concept of vaccination, but it employs other mechanisms to achieve its goals. It would be beneficial to boost the effectiveness of these strategies if we had a clearer understanding of the mechanisms that allow the plant to identify the presence of a virus and the ingenious techniques viruses may employ to avoid the plant's response. There are viruses that cause severe diseases that destroy major staple crops, causing severe economic losses and threatening food supply across the world.

CONCLUSION

CRISPR/Cas is considered as one of the most powerful approaches toward genome editing in various plant species, as this technique is highly efficient, cost effective, and easy to use with low off-target mutations compared to other genome editing techniques like ZFNs and TALENs. CRISPR has been proved as a revolutionary technique in biological research and is becoming the best choice for genome editing due to its various advantages over other GE techniques. Although there are few limitations of this technique such as relatively large size, so as it makes unsuitable to pack into viral vectors, PAM adjustments and off-target effects which are limiting its wide spread use, however, still it is the most robust technique discovered so far. These limitations can be overcome through different strategies such as smaller size CRISPR system, multiple PAM site selection system to increase its efficiency, and recognition ability.

REFERENCES

Abdelkhalek, A., Al-Askar, A. A., Elbeaino, T., Moawad, H., & El-Gendi, H. J. B. (2022). Protective and curative activities of Paenibacillus polymyxa against Zucchini yellow mosaic virus Infestation in Squash Plants. *Biology*, *11*(8), 1150.

Ahmad, A., Ashraf, S., Munawar, N., Jamil, A., Ghaffar, A., & Shahbaz, M. (2021). CRISPR/Cas-mediated abiotic stress tolerance in crops. In *CRISPR Crops*. Springer. pp. 177–211.

Akbar, S., Wei, Y., & Zhang, M (2022). RNA interference: Promising approach to combat plant viruses. *International Journal of Molecular Sciences, 23*(10), 5312.

Anuragi, H., Yadav, R., & Sheoran, R. (2022). Gamma-rays and EMS induced resistance to mungbean yellow mosaic India virus in mungbean [Vigna radiata (L.) R. Wilczek] and its validation using linked molecular markers. *International Journal of Radiation Biology, 98*(1), 69–81.

Balan, M.-T., & Kang, B. (2022). Revisiting Code Normalisation for Machine Learning-based Malware Detection. 45–48.

Banerjee, S., Gupta, S., & Koner, S. (2022). Digital tools for direct and indirect citizen empowerment: The retaliatory response against COVID-19 in India. In *Digital Innovation for Healthcare in COVID-19 Pandemic*. Elsevier. pp. 153–171.

Khan, Z., Khan, S. H., Ahmad, A., Aslam, S., Mubarik, M. S., & Khan, S. (2019). CRISPR/dCas9-mediated inhibition of replication of begomoviruses. *International Journal of Agriculture and Biology, 21*, 711–718.

Binyameen, B., Khan, Z., Khan, S.H., Ahmad, A., Munawar, N., Mubarik, M.S., Riaz, H., Ali, Z., Khan, A.A., Qusmani, A.T., Abd-Elsalam, K.A., Qari, S.H. (2021). Using multiplexed CRISPR/Cas9 for suppression of cotton leaf curl virus. *International Journal of Molecular Sciences, 22*(22), 12543. doi: 10.3390/ijms222212543.

Bánfalvi, Z., Gócza, E., Olasz, F., Pál, M., Posta, K., & Várallyay, É. (2022). "FIBOK 2022": Fiatal Biotechnológusok V. Országos Konferenciája [5th National Conference of Young Biotechnologists]: Program és angol nyelvű összefoglalók [Program and abstracts].

Bhattacharjee, B., & Hallan, V. (2022). Geminivirus-derived vectors as tools for functional genomics. *Frontiers in Microbiology, 13*.

Boro, M., Sannyasi, S., Chettri, D., & Verma, A. K. (2022). Microorganisms in biological control strategies to manage microbial plant pathogens: A review. *Archives of Microbiology, 204*(11), 1–15.

Brown, A. P., Egnin, M., Bukari, F., Idehen, O., Ritte, I., Mortley, D., & Bonsi, C. J. (2022). Cloning and characterization of eukaryotic translation initiation factor 4E (eIF4E) gene family in Ipomoea batatas L.(Lam) for understanding hexaploid sweetpotato-virus interactions. *American Journal of Molecular Biology, 12*(4), 203–244.

Cherusseri, J., Savio, C. M., Khalid, M., Chaudhary, V., Numan, A., Varma, S. J., Kaushik, A. J. B. (2022). SARS-CoV-2-on-chip for long COVID management. *Biosensors, 12*(10), 890.

Di Mattia, J., Torralba, B., Yvon, M., Zeddam, J.-L., Blanc, S., & Michalakis, Y. J. P. (2022). Nonconcomitant host-to-host transmission of multipartite virus genome segments may lead to complete genome reconstitution. *Proceedings of the National Academy of Sciences, 119*(32), e2201453119.

Fan, J., Zheng, J., Lai, W., Liu, S., Liang, X., Wang, Q., Qin, Q. (2023). Establishment of a new intestinal cell line from the grouper (Epinephelus coioides) for aquatic virus pathogenesis. *Aquaculture, 562*, 738771.

Fenibo, E. O., Ijoma, G. N., Nurmahomed, W., & Matambo, T. (2022). The potential and green chemistry attributes of biopesticides for sustainable agriculture. *Sustainability, 14*(21), 14417.

Gao, Z., Zhang, D., Wang, X., Zhang, X., Wen, Z., Zhang, Q., Zhang, Y. (2022). Coat proteins of necroviruses target 14-3-3a to subvert MAPKKKα-mediated antiviral immunity in plants. *Nature communications, 13*(1), 1–20.

Ghattargi, V., Kumari, S. R., Govardhane, O. B., & Pawar, S. P. (2022). Probiotics as a potential immuno modulator to combat coronavirus disease (covid-19) infection. *Egyptian Journal of Medical Human Genetics, 13*(6), 1–4.

Ghosh, S. and Dey, G. (2022). Biotic and abiotic stress tolerance through CRISPR-Cas mediated genome editing. *Journal of Plant Biochemistry and Biotechnology,* 1–12.

Gudeta, W. F., Igori, D., Belete, M. T., Kim, S. E., & Moon, J. S. J. A. o. V. (2022). Complete genome sequence of pueraria virus A, a new member of the genus Caulimovirus. *Archives of Virology, 167*(6), 1481–1485.

Gupta, D. S., Kumar, J., Parihar, A. K., Chandra, A., Sujayanand, G., & Gupta, S. (2022). Urdbean breeding. In *Fundamentals of Field Crop Breeding*. Springer. pp. 1151–1179.

Hillary, V. E., & Ceasar, S. (2022). A Review on the Mechanism and Applications of CRISPR/Cas9/Cas12/Cas13/Cas14 Proteins Utilized for Genome Engineering. *Molecular Biotechnology,* 1–15.

Hinge, V. R., Chavhan, R. L., Kale, S. P., Suprasanna, P., & Kadam, U. (2021). Engineering resistance against viruses in field crops using CRISPR-Cas9. *Molecular Biology Reports, 22*(3), 214.

Hoyos-Villegas, V., Chen, J., Mastrangelo, A. M., & Raman, H. J. (2022). Advances in breeding for quantitative disease resistance. *Frontiers in Plant Science, 13*, 890002.

Ishikawa, M., Yoshida, T., Matsuyama, M., Kouzai, Y., Kano, A., & Ishibashi, K. J. (2022). Tomato brown rugose fruit virus resistance generated by quadruple knockout of homologs of TOBAMOVIRUS MULTIPLICATION1 in tomato. *Plant Physiology, 189*(2), 679–686.

Javed, T., Shabbir, R., Tahir, A., Ahmar, S., Mora-Poblete, F., Razzaq, M., ... Mukhtar, A. (2022). Etiology, epidemiology, and management of maize diseases. In *Cereal Diseases: Nanobiotechnological Approaches for Diagnosis and Management*. Springer. pp. 53–82.

Jones, R. A. (2022). Alteration of plant species mixtures by virus infection: Managed pastures the forgotten dimension. *Plant Pathology, 71*(6), 1255–1281.

Karmakar, A., Taufiqa, S., Baig, M. J., & Molla, K. A. (2022). Increasing disease resistance in host plants through genome editing. *Proceedings of the Indian National Science Academy, 88*(3), 1–13.

Karmakar, S., Das, P., Panda, D., Xie, K., Baig, M. J., & Molla, K. (2022). A detailed landscape of CRISPR-Cas-mediated plant disease and pest management. *Plant Science, 323*, 111376.

Kasanen, R., Awan, H. U. M., Zarsav, A., Sun, H., & Asiegbu, F. O. (2022). Forest tree disease control and management. In *Forest Microbiology*. Elsevier. pp. 425–462.

Kaur, P., Kaila, T., Dhkal, M., & Gaikwad, K. (2022). Chloroplast genome and plant–virus interaction. In *Bioinformatics in Agriculture*. Elsevier. pp. 419–436.

Kerber-Diaz, J. C., Leos-Ramírez, M. A., Flores-Ceron, A. A., Ponce-Mendoza, A., Estrada-de los Santos, P., & Ibarra (2022). Distribution of CRISPR-Cas systems in the Burkholderiaceae family and its biological implications. *Archives of Microbiology, 204*(12), 1–12.

Khan, Z., Saboor, T., Ashfaq, M., Saddique, A., & Khanum, P. (2021). Disease resistance in crops through CRISPR/Cas. In *CRISPR Crops*. Springer. pp. 151–175.

Khan, Z. A., Kumar, R., & Dasgupta, I. (2022). CRISPR/Cas-mediated resistance against viruses in plants. *International Journal of Molecular Sciences, 23*(4), 2303.

Klymiuk, V., Chawla, H. S., Wiebe, K., Ens, J., Fatiukha, A., Govta, L., ... Pozniak, C. (2022). Discovery of stripe rust resistance with incomplete dominance in wild emmer wheat using bulked segregant analysis sequencing. *Communications Biology, 5*(1), 1–9.

Kulabhusan, P. K., Tripathi, A., & Kant, K. (2022). Gold nanoparticles and plant pathogens: An overview and prospective for biosensing in forestry. *Sensors, 22*(3), 1259.

LaTourrette, K., & Garcia-Ruiz, H. (2022). Determinants of virus variation, evolution, and host adaptation. *Pathogens, 11*(9), 1039.

Liu, L., & Pei, D. (2022). Insights gained from RNA editing targeted by the CRISPR-Cas13 family. *International Journal of Molecular Sciences, 23*(19), 11400.

Lopez-Gomollon, S., & Baulcombe, D. C. (2022). Roles of RNA silencing in viral and non-viral plant immunity and in the crosstalk between disease resistance systems. *Reviews Molecular Cell Biology, 23*(10), 645–662.

Ma, Y., Xing, F., Che, H., Gao, S., Lin, Y., & Li, S. (2022). The virome of Piper nigrum: Identification, genomic characterization, prevalence, and transmission of three new viruses of black pepper in China. *Plant Disease, 106*(8), 2082–2089.

Mahas, A., & Mahfouz, M. (2018). Engineering virus resistance via CRISPR–Cas systems. *Current Opinion in Virology, 32*, 1–8.

Malik, I. M., Tak, H., Lone, G., Dass, W. M. (2022). Phytoparasitic nematodes as the major threat to viticulture. *Environmental and Experimental Biology, 20*(1), 1–10–11–10.

Mangukia, N., Rao, P., Patel, K., Pandya, H., Rawal, R. M. (2022). Unveiling the nature's fruit basket to computationally identify Citrus sinensis csi-mir169–3p as a probable plant miRNA against Reference and Omicron SARS-CoV-2 genome. *Computers in Biology and Medicine,* 105502.

Martínez-Alonso, E., Guerra-Pérez, N., Escobar-Peso, A., Peracho, L., Vera-Lechuga, R., Cruz-Culebras, A., Alcázar, A. (2022). Phosphorylation of eukaryotic initiation factor 4G1 (eIF4G1) at Ser1147 is specific for eIF4G1 bound to eIF4E in delayed neuronal death after ischemia. *International Journal of Molecular Sciences, 23*(3), 1830.

Mubarik, M. S., Wang, X., Khan, S. H., Ahmad, A., Khan, Z., Amjid, M. W., Food. (2021). Engineering broad-spectrum resistance to cotton leaf curl disease by CRISPR-Cas9 based multiplex editing in plants. *GM Crops & Food, 12*(2), 647–658.

Osborne, M. J., Volpon, L., Memarpoor-Yazdi, M., Pillay, S., Thambipillai, A., Czarnota, S., Cowling, V. H. (2022). Identification and characterization of the interaction between the methyl-7-guanosine cap maturation enzyme RNMT and the cap-binding protein eIF4E. *Journal of Molecular Biology, 434*(5), 167451.

Pandita, D. (2022). CRISPR/Cas-mediated genome editing technologies in plants. In *Plant Abiotic Stress Physiology*. Apple Academic Press. pp. 293–330.

Patra, G. K., Gupta, D., Rout, G. R., Panda, S. (2022). Role of long non coding RNA in plants under abiotic and biotic stresses. *Plant Physiology and Biochemistry, 194*, 96–110.

Payne, S. (2022). *Viruses: From Understanding to Investigation*: Elsevier.

Pires, D., Vicente, C. S., Menéndez, E., Faria, J. M., Rusinque, L., Camacho, M. J., & Inácio, M. (2022). The Fight against Plant-Parasitic Nematodes: Current Status of Bacterial and Fungal Biocontrol Agents. *Pathogens, 11*(10), 1178.

Plchová, H., Moravec, T., Čeřovská, N., Pobořilová, Z., Dušek, J., Kratochvílová, K., Kundu, (2022). A Goldenbraid-compatible virus-based vector system for transient expression of heterologous proteins in plants. *Viruses, 14*(5), 1099.

Popoola, T. D., Segun, P. A., Ekuadzi, E., Dickson, R. A., Awotona, O. R., Nahar, L., Fatokun, A. A. (2022). West African medicinal plants and their constituent compounds as treatments for viral infections, including SARS-CoV-2/COVID-19. *DARU Journal of Pharmaceutical Sciences,* 1–20.

Prasher, P., & Sharma, M. (2022). An Introduction to Rice Diseases. In *Cereal Diseases: Nanobiotechnological Approaches for Diagnosis and Management.* Springer. pp. 3–15.

Qin, S., Xiao, W., Zhou, C., Pu, Q., Deng, X., Lan, L., Therapy, T. (2022). Pseudomonas aeruginosa: Pathogenesis, virulence factors, antibiotic resistance, interaction with host, technology advances and emerging therapeutics. *Signal Transduction and Targeted Therapy, 7*(1), 1–27.

Rahman, S. U., McCoy, E., Raza, G., Ali, Z., Mansoor, S., & Amin, I. (2022). Improvement of Soybean; A Way Forward Transition from Genetic Engineering to New Plant Breeding Technologies. *Molecular Biotechnology,* 1–19.

Rai, A., Sivalingam, P. N., & Senthil-Kumar, M. (2022). A spotlight on non-host resistance to plant viruses. *PeerJ, 10*, e12996.

Rubiales, D., Khazaei, H.., & Genetics, A. (2022). Advances in disease and pest resistance in faba bean. *Theoretical and Applied Genetics*, 1–22.

Saberi R., Gholizadeh Vazvani, M., Ebrahimi-Zarandi, M., & Skorik, Y. (2022). Alginate-induced disease resistance in plants. *Polymers, 14*(4), 661.

Saikia, R., Paswan, R. R., Farwaha, N., Borah, B. K., & Decade, M. (2022). Plant Viruses: Factors Involved in Emergence and Recent Advances in Their Management. *Challenges and Management in the New Decade,* 29–55.

Severin, R. (2022). *Veterinary virology. Part 1. General Virology*. Faculty of Veterinary Medicine, Department of Epizootology and Micrjbiology, State Biotechnological University.

Shahriar, S. A., Islam, M. N., Chun, C. N. W., Rahim, M. A., Paul, N. C., Uddain, J., & Siddiquee, S. (2021). Control of plant viral diseases by CRISPR/Cas9: resistance mechanisms, strategies and challenges in food crops. *Plants, 10*(7), 1264.

Sharma, N., Prasad, A., & Prasad, M. (2022). Role of the Sw5 gene cluster in the fight against plant viruses. *Journal of Virology, 96*(5), e02084–02021.

Shukla, A., Hoffmann, G., Kushwaha, N. K., López-González, S., Hofius, D., & Hafrén, A. (2022). Salicylic acid and the viral virulence factor 2b regulate the divergent roles of autophagy during cucumber mosaic virus infection. *Autophagy, 18*(6), 1450–1462.

Tiwari, J. K., Tuteja, N., & Khurana, S. (2022). Genome editing (Crispr-Cas)-mediated virus resistance in potato (*Solanum tuberosum* L.). *Molecular Biology Reports*, 1–11.

Tripathi, L., Ntui, V. O., Tripathi, J. N., & Kumar, P. (2021). Application of CRISPR/Cas for diagnosis and management of viral diseases of banana. *Frontiers in Microbiology, 11*, 609784.

Tuncel, A., & Qi, Y. (2022). CRISPR/Cas mediated genome editing in potato: Past achievements and future directions. *Plant Science*, 111474.

Tyagi, S., Kumar, R., Kumar, V., Won, S. Y., Shukla, P. & Food. (2021). Engineering disease resistant plants through CRISPR-Cas9 technology. *GM Crops & Food, 12*(1), 125–144.

Van Rythoven, E. J. (2022). Backstage mockery: Impoliteness and asymmetry on the world stage. *Global Studies Quarterly, 2*(4), ksac064.

Varanda, C. M., Félix, M. d. R., Campos, M. D., Patanita, M., & Materatski, P. (2021). Plant viruses: From targets to tools for CRISPR. *Viruses, 13*(1), 141.

Voloudakis, A. E., Kaldis, A., & Patil, B. L. (2022). RNA-based vaccination of plants for control of viruses. *Annual Review of Virology, 9*, 521–548.

Wang, P., Liu, J., Lyu, Y., Huang, Z., Zhang, X., Sun, B., Zhang, C. (2022). A review of vector-borne rice viruses. *Viruses, 14*(10), 2258.

Wani, S. H., Samantara, K., Razzaq, A., Kakani, G., & Kumar, P. J. (2022). Back to the wild: Mining maize (Zea mays L.) disease resistance using advanced breeding tools. *Molecular Biology Reports*, 1–17.

Watson, B. N., Staals, R. H., & Fineran, P. C. (2018). CRISPR-Cas-mediated phage resistance enhances horizontal gene transfer by transduction. *MBio, 9*(1), e02406–02417.

Wiebe, S., Huang, Z., Skalecka, A., Cagnetta, R., Lacaille, J.-C., Aguilar-Valles, A., & Sonenberg, N. (2022). Cell-type-specific translational control of spatial working memory by the cap-binding protein 4EHP. *Molecular Brain, 16*(1), 1–13.

Wu, W., Luo, X., Ren, M. (2022). Clearance or Hijack: Universal Interplay Mechanisms Between Viruses and Host Autophagy From Plants to Animals. *Frontiers in Cellular and Infection Microbiology*, 1345.

Wulff, B. B., & Krattinger, S. G. (2022). The long road to engineering durable disease resistance in wheat. *Frontiers in Genetics, 73*, 270–275.

Yadav, M. K., Patel, P., Kumar, A., Modi, A., Babu, P. J., Chhakchhuak, L.,Singh, P. K. (2022). Tailoring Disease Resilience Crops through CRISPR/Cas. In *Microbial Biocontrol: Food Security and Post Harvest Management*. Springer. pp. 187–209.

Yang, M., Ismayil, A., Jiang, Z., Wang, Y., Zheng, X., Yan, L., Liu, Y. (2022). A viral protein disrupts vacuolar acidification to facilitate virus infection in plants. *The EMBO Journal, 41*(2), e108713.

Yang, M., & Liu, Y. J. F. l. (2022). Autophagy in plant viral infection. *Biomedicine & Pharmacotherapy, 158*, 114086.

Yao, S., Kang, J., Guo, G., Yang, Z., Huang, Y., Lan, Y., Xu, Z. (2022). The key micronutrient copper orchestrates broad-spectrum virus resistance in rice. *Science Advances*, *8*(26), eabm0660.

You, H., Gordon, C. A., MacGregor, S. R., Cai, P., & McManus, D. (2022). Potential of the CRISPR-Cas system for improved parasite diagnosis: CRISPR-Cas mediated diagnosis in parasitic infections. *BioEssays*, *44*(4), 2100286.

Yue, J., Wei, Y., & Zhao, M. (2022). The reversible methylation of m6A is involved in plant virus infection. *Biology*, *11*(2), 271.

Zhang, T., Shi, C., Hu, H., Zhang, Z., Wang, Z., Chen, Z., Lu, Q. (2022). N6-methyladenosine RNA modification promotes viral genomic RNA stability and infection. *Theranostics*, *13*(1), 1–16.

Zhou, S. & Liu, Y. (2022). A study of pragmatic vagueness in Chinese spokespersons' remarks. *Journal of Literature and Art Studies*, *12*(5), 469–475.

6 Utilizing CRISPR-Cas System to Develop Resistance against Bacterial Diseases

INTRODUCTION

All plants have defense mechanisms, but having resistant genes is not a rule. Still, most of the bacterial pathogens caused diseases in a number of plants under particular conditions. The mechanism that regulates the plant–bacterial pathogen interactions will be leading toward the disease and disease resistance. Molecular biology and biotechnology have contributed to reveal these mechanisms through modern techniques. Most of the plant–bacteria interactions are highly specific. A series of molecular recognition processes have been involved between the plant and bacterial pathogens. The exchange of information between the two partners and suitable events induced by the pathogen attack determines that the interaction will lead to the disease. Some examples of bacterial diseases are given on Table 6.1.

Plants have some molecules that derive functions as a signal that induces the expression of genes of bacteria during the interaction between them. Ultimately, the products of these genes induce changes in the expression of plant genes that result in the production of metabolites in plants. This type of interaction has been seen in almost all bacterial pathogens, irrespective of compatible or incompatible interactions. A number of chemicals, physical and biological methods have been used to protect crops against bacterial diseases. But some diseases have not been controlled due to changing climatic conditions and development of resistance in bacterial pathogens.

To engineer plant resistance to bacterial diseases, a variety of genetic strategies have been proposed, bacterial pathogenicity or virulence factors, producing antibacterial proteins of non-plant origin enhancing natural plant defense and artificially inducing programmed cell death at the site of infection. These are dependent on the knowledge of the mechanism of actions and successive steps on plant–bacteria interaction. CRISPR-Cas has been utilized excessively in plants, bacteria and other organisms for targeting specific genes.

With a paradigm shift, the use of the genome editing (GenEd) tool for targeted genome alterations in various species has transformed the landscape of life sciences. CRISPR-Cas9 has been used to make targeted genomic changes. The GenEd tools can find and bind certain DNA sequences, which may then be used to target any sequence of interest. The CRISPR system was driven by a microbial adaptive

DOI: 10.1201/b22901-6

immune response that uses gRNA-nuclease to recognize and cut foreign DNA segments (Jinek *et al.* 2012).

DURABLE MANAGEMENT OF BACTERIAL DISEASE AND PLANT HOST RESISTANCE

The defensive mechanism in plants begins with cell-to-cell contact between the host plant and a bacterial pathogen. Several compounds from bacterial pathogens and plants are transferred or interact in the course of pathogen–plant interaction, which is a complicated series of processes (Silva *et al.* 2018). The sort of biomolecules used for defense is mostly determined by the pathogen, such as bacteria (Silva *et al.* 2018). Bacteria, for example, secretes virulence-associated proteins through type II, III and IV secretion systems (Kamber *et al.* 2017) and work together with host plants. All these contagious biomolecules cause the plant to get infected or develop haustorium (Kamoun, 2006), and have a detrimental impact on its health.

The intricate connections that exist between innate immunity in plants, gene regulation and ETS-associated resistance, in the context of plant–pathogen interactions. PTI (PAMPs-triggered immunity), also known as ETI (effector-triggered immunity), is a type of immunity that is induced in response to a bacterial pathogen attack by receptors localized in plasma membrane known as PRR that recognize the presence of pathogen-associated molecular patterns or effectors in the extracellular environment. Bacteria belongs to the 'ETI' type of pathogen. The term 'effector' refers to any regulatory molecule that is secreted by a pathogen and then modifies a host protein in such a way as to establish the pathogen's growth and cause ETS (effector-triggered susceptibility). R proteins are the proteins which enables the plant to resist the onslaught of pathogens.

Receptors recognized as a trans-membrane protein are some kinds of receptors found in plants (Kachroo *et al.* 2017). These receptors detect bacterial pathogen-linked molecular patterns (MAMPs and PAMPs) and launch a defense response recognized as a PTI to combat the contagious pathogen (Andersen *et al.* 2018; Tyagi *et al.* 2018). Some bacterial pathogens release effectors in response to this defensive mechanism to resist plant immunity, by activating susceptibility proteins to decrease PTI and speed up infection process (Kachroo *et al.* 2017). As a result, to combat effectors, plants activate *R* genes. Signals triggered the related genes from the *Avr* proteins or effectors that will be resulting in ETI.

PAMPs-triggered immunity is the common type of immunity that is present among a group of pathogens and triggered by the identification of PAMPs. On the other hand, ETI is a sort of immunity, and it is stimulated by pathogen effector recognition, involves PCD via the hypersensitive response and regulates proliferation of the pathogen at the site of the infection (Oskar *et al.* 2015). ETI and PTI stimulate the expression of pathogenesis-related genes, change mitogen-activated protein kinase (MAPKs), which are derived from kinases and affect transcription factors or hormones of plants in response to pathogen infection. It also influences posterior events such as the production of reactive oxygen species, HR, the release of antimicrobial proteins, stomata closure, cell wall modification and chemicals like phytoalexins chitinases, defensins and protease inhibitors. Plants utilized RNA interference method to detect or remove certain infections of viruses, in addition to PTI and ETI (Dong & Ronald, 2019; Rosa *et al.* 2018).

COMPLEXITIES RELATED TO BACTERIAL DISEASE RESISTANCE IN PLANTS

Plants and pathogens are always fighting to protect or destroy each other in nature. As a result, plants must employ a variety of defensive measures to combat bacterial pathogens. Plants evolve diverse defensive mechanisms and pathogens with superior escape skills as a result of selective pressure in the zig-zag model, outlined by Birch and Prichard (Pritchard & Birch, 2014). Plants, for example, employ plasma membrane and inter-/intracellular sensors in addition to physical barriers (which restrict pathogen entrance into the cell) to begin protection methods, if the pathogen or pathogen-derived alterations in host cells are detected (Andersen *et al.* 2018). Plants produce antimicrobial agents that detect pathogen-derived chemicals or pathogens and kill them either by inhibiting its virulence or detoxification (Mushtaq *et al.* 2019).

Pathogens have developed to defeat plant defense mechanisms and gain control of the plant system to begin and propagate infection. Several bacteria release cell wall-degrading enzymes like xylanase and cellulose, which may penetrate in cell membrane of plants and allow these pathogens to infiltrate the host (Kubicek *et al.* 2014). Some pathogens secret effector molecules to contribute to disease progression and to suppress defense system of host (Aman *et al.* 2018). Understanding of these molecular interactions between plants and pathogens has allowed genetic engineers to create less sensitive and disease-resistant plant for the improvement of agriculture (Kumar *et al.* 2019).

Molecular biology developments have made it easier to edit a host's genome by adding genes that can excrete antimicrobial compounds, break down toxins and inhibit cell wall-degrading enzymes or by removing pathogen-prone genes (Kumar *et al.* 2015). Advanced molecular technologies make it easier to investigate infection and immunology in plants and pathogens and edit their genomes to generate plant infection resistance. CRISPR-centered GE is used to modify plant traits and increase disease resistance.

METHODS OF CRISPR TO DEVELOP RESISTANCE AGAINST BACTERIAL DISEASE IN PLANTS

The CRISPR-Cas system, which was generated by *Streptococcus pyrogens*' adaptive immune system, has encouraged the modifications of the genes in a number of crop species. CRISPR-Cas is an important or potent tool of crop genome editing, as it is easy to design, has an efficiency and edits numerous genes easily (Tripathi *et al.* 2021).

CRISPR-Cas9 is made up of Cas9 nuclease and synthetic guided RNA (sgRNA). By comparing 50 leading sequences of the sgRNA and the DNA, Cas9 can identify the target DNA. The PAM (protospacer adjacent motif), a three-nucleotide sequence region that mainly consists of NAG or NGG, where N is any nucleotide, is recognized, permitting Cas9 to initiate comparing upstream sequence to gRNA. Three to four nucleotides are generated in the downstream region from the cut site of Cas9. Spacer sequence of 20 nucleotides and a scaffold of sgRNA is involved to target the genome sequence. It prompts Cas9 to produce specific double-strand breaks (DSBs) that are exact and accurate. If the donor template is available, the DSBs are repaired by NHEJ or HDR, leading to small insertions, deletions, gene replacement or gene substitution.

To develop resistance against bacterial diseases in plants, apart from Cas9, other Cas proteins have been used. Class 2, type V-CRISPR Cas12a (cpf1) only possesses RuvC domains. It has single-strand DNase and crRNA biogenesis RNase activity, and it can be identified the T-rich PAM. Cas12 might be utilized for multiplex gene editing that used a single sequence array on the designated sgRNA. ssRNA molecules of the phage gene may be targeted and cut by Cas13a, a class 2 type V1-A ribonuclease (Aman *et al.* 2018). Since it is more precise than PCR-detecting viruses, it may be utilized by RNA viruses. There is no need of PAM segment for Cas13a.

Three different gene editing systems, known as SDN1, SDN2 and SDN3, have been discovered based on the repair process (Modrzejewski *et al.* 2018). SDN1, which is based on NHEJ, is a very effective yet error-prone repair of a targeted DSB. The host genome undergoes unpredictable modifications as a result of the DSB repair, which might lead to gene silencing, knockouts or altered function. SDN2, which has a repair template that has sequence match to the target site added to the CRISPR-Cas reagent, is less effective and higher fidelity. Following the HDR repair of the DSB in the SDN2, nucleotide replacement or targeted insertions or deletions are produced. High fidelity but even less effective enzymatic activity can be seen in SDN3. The whole gene or genetic element(s) are inserted at the target site depending on donor sequence, after the DSB in SDN3 is repaired using HDR using the donor template. Base editing (BE), another sort of editing system, is utilized for target editing of a single base pair.

To create a base editor that enables single-nucleotide base replacement resolution without a DNA donor template, DNA deaminase and dCas9 must be fused. Depending on the DNA deaminases utilized, the DNA deaminases work as effectors, allowing C:G-to-T:A or A:T-to-G:C substitution, while the RNA-guided CRISPR system acts as a genomic locator of the targeted area.

Prime editing, a new SDN editing tool, was just created. Prime editing mediates DNA base pair swaps, modest insertions and small deletions using the exact same mechanism as traditional CRISPR-Cas systems. Primer editing, on the other hand, decreases off-target effects and addresses frameshifts brought on by indels without causing DSB or requiring a donor template (Chen *et al.* 2021). Prime editing calls for a pegRNA, a longer-than-normal sgRNA and a fusion protein made of Cas9 H840A nickase linked to a designed RT enzyme. While prime editing has the potential to supplement the current CRISPR editing technologies and enable precise and targeted DNA alterations, its biological determinants are still poorly understood.

Prime editing is a fascinating method that may be used for banana gene editing, though. Base editing and prime editing may be regarded as SDN1 and SDN2 kinds of gene editing, respectively, as they do not require a DNA donor template (Tripathi *et al.* 2021). As a result, they may be recognized as nontransgenic products and do not need to follow the same biosafety requirements as transgenic products. But above all the most and favorable tool used to combat bacterial diseases in plants is CRISPR-Cas9. It can easily be used for developing resistance in majority of the plants against bacterial pathogens.

CRISPR-CA9: A BACTERIAL IMMUNE SYSTEM AND AN IMPORTANT PLANT EDITING TOOL

An RNA-mediated immune system that is used to protect prokaryotes from bacteriophages and viruses is called CRISPR-Cas9 (Hu *et al.* 2019). The two components of this system are referred to as the CRISPR array and Cas9 nucleases. The first component CRISPR array is a set of repetitions of 50 base pairs (bp) that are separated by spacers of identical length and individuality. Contrarily, the protein Cas9, which is associated with CRISPR, helps to degrade external DNA by acting as a nuclease (Makarova *et al.* 2015; Ricroch *et al.* 2017). When the pathogen attack, genetic material from outside virus is found and incorporated into array. This experience causes the bacterial genome to create a form of memory. The CRISPR array is then transcribed into mature crRNA. CrRNA biogenesis is the name given to this process. This crRNA also identifies and degrades viral DNA sequence that includes complimentary sequence to the crRNA (Ahmad *et al.*, 2020). The Cas9 nuclease aids in DNA degradation (Bhaya *et al.* 2011; Liu *et al.* 2016).

Because of its ability to cause double-strand breaks in the genome, the CRISPR-Cas9 system is well suited for use as genomic scissors in the process of genetic editing (Jinek *et al.* 2012). It does not require any debonair apparatus to be carried out in the lab apart from (1) a guided RNA sequence that is 18–22 bp long and corresponding to the direct target site and (2) Cas9 nuclease to provoke a double-strand break after the sequence of PAM (Chen *et al.* 2019).

Two approaches are used to repair CRISPR induced DSB: one is NHEJ (non-homologous end joining), and the other is HDR (homology-directed repair) (Liu *et al.* 2019). Because of the error-prone nature of NHEJ, it has the potential to cut down the target sequence by adding deletions and insertions, which may lead to the loss of a function as well as the development of genetically modified plants. After the synthesis of gRNA and the identification of the PAM sequence, the gRNA is cloned into the selected vector. This is followed by transformation, screening and confirmation (Lone *et al.* 2018; Pickar-Oliver & Gersbach, 2019). CRISPR-Cas9 has been used effectively to change the genomes of many plant species, ranging from monocots to dicots, as well as the genomes of almost every other living system. Regardless of their genomic intricacy in terms of ploidy level, the genome of cash crops such as soybean, citrus, rice, barely, cotton, potato and tomato as well as model plants such as *Nicotiana benthamiana* and *Arabidopsis thaliana* have been changed (Brooks *et al.* 2014; Fauser *et al.* 2014). **Infection**: The viral genome injects into the host, after infection. **Adaptation**: CRISPR array (spacers and repetitions) is utilizing Cas proteins to get the viral genome. **crRNA biogenesis**: RNA polymerase transcribes Pre-crRNA from the larger region, and then splits into minor crRNAs with a single spacer and repetition is restricted by Cas proteins. **Interference**: The presence of a crRNA spacer that is highly complementary to the invading viral genome is required for the initiation of a cleavage (Figure 6.1). Cas proteins are essential to the mechanism that impedes the replication of viruses and need their presence. This is how CRISPR-Cas9 functions as a bacterial immune system.

The following steps are typically accompanied when altering the genome of plants: firstly, to find out the target gene and producing sgRNA according to the

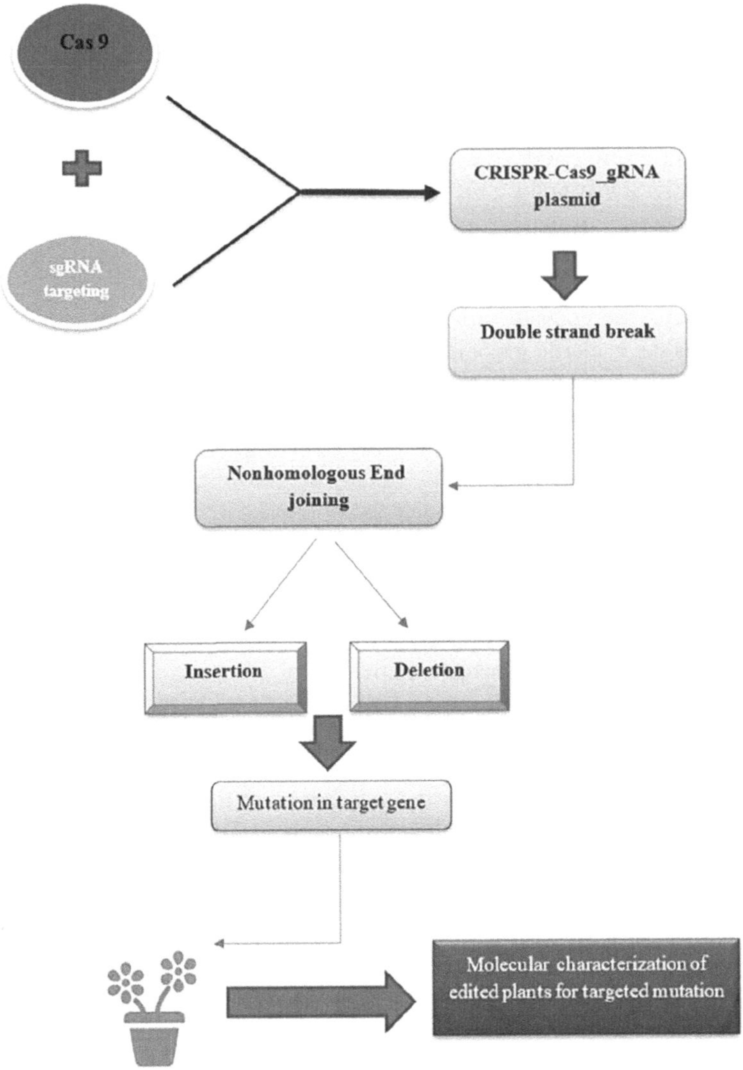

FIGURE 6.1 CRISPR defense system in bacteria.

target gene. Plant vector will be constructed and finally performed plant transformation. The single-guided RNA leads Cas9 protein to compel the target sequence and create double-strand break after transformation. Random mutations developed by DSB and repaired by HDR is a brief explanation for CRISPR that has been proved by some experimental work. The mutations would be detected with the help of computational technologies. Cas stands for CRISPR-associated, DSB stands for double-strand break, HDR stands for homology-directed repair, NHEJ stands for non-homologous end-joining and sgRNA stands for single guide RNA (Table 6.1).

TABLE 6.1
Bacterial Diseases in Plants

Target Function	TALE	Gene Name	Species	Symptoms	Host	Effect	Reference
Transcription factor	PthXo6	*TFX*-1	*Xoo*	Water soaking	Rice	Increased growth and lesions	Sugio *et al.* (2007)
	PthXo7	*TFII*Aγl					Hu *et al.* (2014)
	AvrBs3	*UPA*-20	*Xav*	Cell hypertrophy	Pepper	Increased leaf area	Kay *et al.* (2009)
	PthA	*CsL*OB-1	*Xcc*	Cell expansion	Citrus	Canker	Hu *et al.* (2014)
	PthB		*Xca*				Al-Saadi *et al.* (2007)
Stability of miRNA		*SlD*MR6-1	*Xanthomonas* spp.	Regulation of pathogen responsive gene	Tomato		Thomazella *et al.* (2016)
	PthXo8	*OsH*EN-1	*Xoo*	Soaking of water	Rice	Not reported	Moscou and Bogdanove (2009)
Sugar Transporter	PthXo-1	*Os*SWEET 11	*Xoo*	Water soaking	Rice	Increased growth and lesions	Chu *et al.* (2006)
							Yang and White (2004)
	PthXo-2	*Os*SWEET13					Zhou *et al.* (2015)
	PthXo-3	*Os*SWEET14					Yang and White (2004)
	Tal5						Streubel *et al.* (2013)
	TalC						Yu *et al.* (2011)
	AvrXo7						Antony *et al.* (2010)
	PthA series	*Cs*SWEET1	*Xoc*	Unknown	Citrus	Unknown	Jia *et al.* (2016)
	Tal20	*Me*SWEET 10a	*Xam*	Water soaking	Cassava	Unknown	Cohn *et al.* (2014)
Sulfate transporter	Tal2g	*SUL*TR3:6	*Xoc*	Water soaking	Rice	Increased lesions on leaf surface	Cernadas *et al.* (2014)

EDITING HOST SUSCEPTIBILITY GENES TO ENHANCE BACTERIAL RESISTANCE

Bacterial pathogens are diverse, ubiquitous, multiply rapidly and perform an important (beneficial or destructive) function in the biological system (Yin & Qiu, 2019). Phytopathogenic bacteria are a huge concern to agriculture, because they produce a variety of diseases like rots, mosaics, or spots, and they are difficult to cope, resulting in considerable production loss (Vale *et al.* 2001; Zeng *et al.* 2010). Crop-specific bacteria create distinct disease symptoms in certain hosts such as *Clavibacter michiganensis* causes tomato bacterial ring rot disease. *Ralstonia solanacearum*, on the other hand, generates disease symptoms and has an extensive host range, involving monocots and dicots. Kingdom crossers, such as *Dickeya dadantii*, are adaptable pathogens that may trigger a variety of diseases in animals and plants. Pathogens of various kinds elicit varied reactions in their hosts, making resistance development difficult. Bacteria acquire resistance to antibacterial agents and transmit infection as a result of their constantly changing nature and horizontal gene transfer (Borrelli *et al.* 2018).

Bacteria enter the cell by natural mechanisms such as stigma, stomata and other plant openings, or artificial wounds, or by secreting bioactive chemicals generally (Zeng *et al.* 2010). These signals are recognized by the plant defense system, which responds properly to eradicate infection. Successful pathogens, on the other hand, employ a complicated signal cascade in which many host plant genes, including several *S* genes, play a role, allowing pathogens to overcome plant protection and commence infection (Dong & Ronald, 2019; Dominguez *et al.* 2016). Because the resistance provided by the *S* gene could be more long lasting, as was mentioned before, both the genes are prime candidates for genome editing in order to offer bacterial disease resistance. However, the production of cultivars that are resistant to bacterial diseases has been going at a slower rate. The reasons for this are the genetic diversity of the plants and the continual growth of bacteria due to their ability to evade the protective systems of plants. Despite this, minimum study has been conducted using the CRISPR-Cas9 system to grow transgenic plants resistant to bacterial diseases as compared to viral and fungal diseases.

The bacterium *Xanthomonas citri* causes citrus canker disease, which is an economically important bacterial pathogen that causes large yield loss in citrus plants (Peng *et al.* 2017). Editing the genes involved in ETI to build the resistance against *X. citri*, CRISPR-Cas9 method was used. Promoter region of the *CsLOB-1* (lateral organ boundaries 1) gene, the effectors binding element (*PthA4*), which was implicated in host susceptibility, was altered, and as a consequence, the ability to locate and react to bacterial effectors was lost, resulting in greater resistance to infection. Jia *et al.* (2022) first modified the *EBEPthA4* element in *CsLOB1*'s promoter region. They claimed that the symptoms of disease were decreased without changing plant phenotypes, and that no off-target mutations were produced (Peng *et al.* 2017). This research expanded on the previous findings by demonstrating a relationship between the *CsLOB-1* promoter and formation of disease in citrus plants and concluding that deleting *EBEPthA4* enhanced resistance to *X. citri* with no phenotypic changes.

Bacterial pathogens with various species like Xanthomonas and *Pseudomonas syringae* affected tomato plants, declining tomato yield and inflicting serious damage to the economy, were also edited by using CRISPR-Cas9 techniques (Schwartz *et al.* 2015). The expression of *DMR6* was enhanced by *Pseudomonas syringae pv. Tomato DC3000*-infected tomato, which acts like a negative regulator to plant immunity as well as aids in infection dissemination (Langner, Kamoun, & Belhaj, 2018; Zeilmaker *et al.* 2015). However, obliteration of tomato orthologue '*SlDMR6*' through CRISPR-Cas9 increased tomato plant resistance to several pathogens, including *Phytophthora capsica* and *P. syringae* pv tomato. When the *DMR*6 gene in *A. thaliana* plants was edited by CRISPR-Cas9, it raised the amount of salicylic acid which is a plant hormone involved in plant defense and offered resistance against plant diseases. A general framework of developing disease-resistant crops using CRISPR-Cas9 system is given in Figure 6.2.

The findings of these research support the use of genome editing to inactivate and delete specific genes and provide resistance to large spectrum of plant diseases. In a separate study, CRISPR technology was utilized to deliver resistance against bacterial blight pathogen *Xanthomonas oryzae* via modifying the *OsSWEET13* gene, a vulnerable host susceptibility gene implicated in the transfer of sucrose during pathogenesis (Zhou, Liu, Weeks, Spalding, & Yang, 2014). The *X. oryzae* effector protein *PthXo2*, which boosts host susceptibility, induces the expression of *OsS*WEET13. Susceptibility gene *OsS*WEET13 in rice knocked out by Zhou, resulted in a mutation that caused resistance against *X. oryzae.*

CRISPR-Cas9 are used to produce *JAZ2* mutations that restrict stomata opening and confer resistance to bacterial speck disease (Ortigosa *et al.*, 2019). Resistance to biotrophic pathogens usually results in vulnerability to necrotrophic pathogens and vice versa (Gimenez-Ibanez *et al.* 2017). Signals connected to jasmonic acid defense system stayed out of stomata in this situation; hence, they had no influence on the SA defense system.

Thus, *Botrytis cinere* a necrotrophic pathogen, which causes grey mold disease in tomatoes, had no effect on *Sl*jaz21jas mutant plants. This research presents an effective model for removing both (JA and SA) defensive mechanisms and a concept for targeting the host genome to provide resistance to a wide spectrum of infections. *DspE* is a pathogen effector protein produced by *Erwinia amylovora.* This interaction between the *DspE* protein and the *DspE*-interacting proteins of Malus genes *(DIPM)*, which include *DIPM* one, two, three and four, causes plants to be vulnerable to fire blight (Das *et al.* 2019). With the use of CRISPR-Cas9, these *DIPM* genes were silenced in apple's protoplasts so that the fruit would be resistant to the fire blight disease (Malnoy *et al.* 2016). Research indicates that *S* gene editing and CRISPR-Cas9 are useful tools for developing resistance not just in plants whose life cycles are relatively brief but also in trees, which have a much longer lifetime. *Evo*lvR, a modified version of CRISPR-Cas9 that can recognize *Xanthomonas* strains and their ligands, was used to develop unique alleles in rice (Luu *et al.* 2019). Long-term farming may benefit from the use of transgenic crops, since these crops include distinct

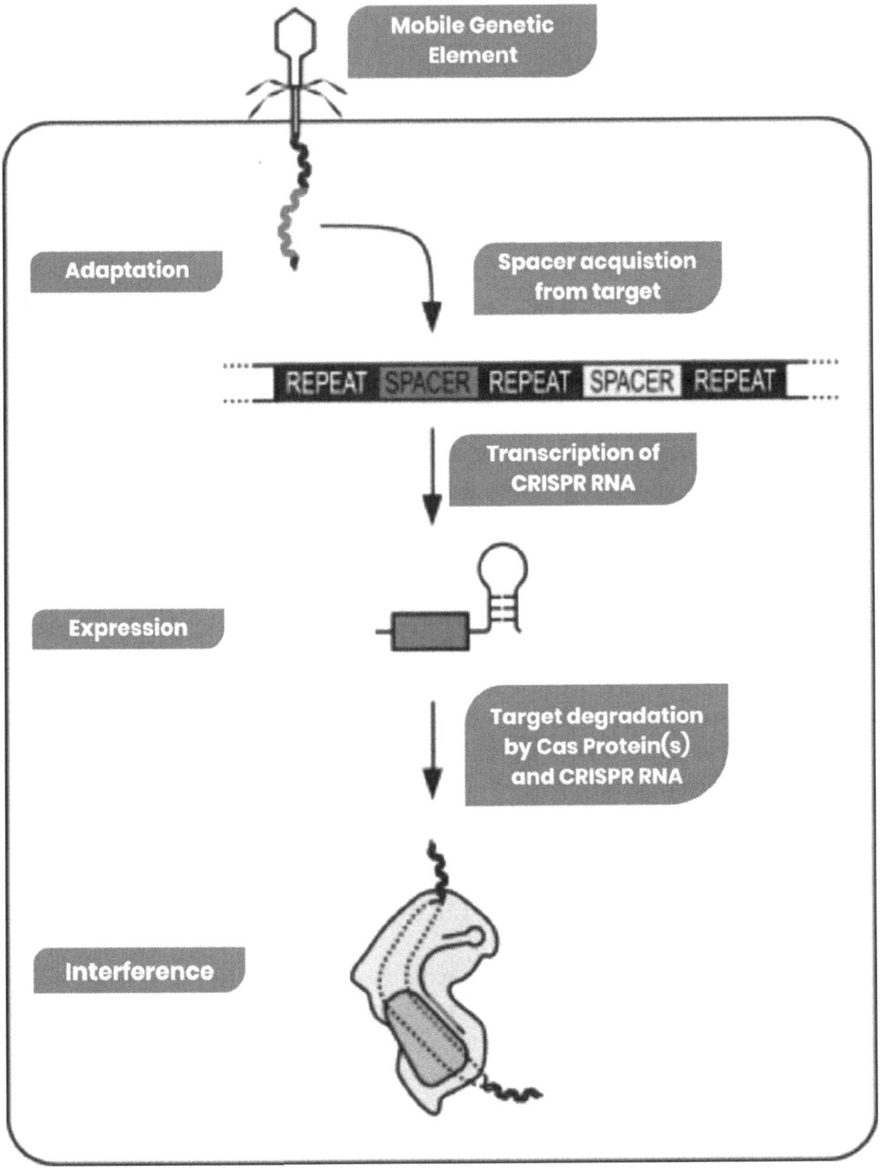

FIGURE 6.2 General framework of developing disease-resistance crops using CRISPR-Cas9 system.

genes that provide resistance against infections. In the future, development of more accurate technologies that can quickly and easily diagnose various diseases and build resistance in plants will widen the application of disease resistance in plants and accelerate the evolution of disease resistance in plants. Examples of CRISPR edited plants are given in Table 6.2.

TABLE 6.2
Edited Plants to Build Up Resistance against Pathogen

Bacterial Pathogen	Disease Caused	Name of Plant	Target Gene	Location	Function	Method Used	Successful Reports	References
Xanthomonas species	Bacterial blight	*Solanum Lycopersicom*	*SlDMR6*	Host *S* gene	Pathogen responsive gene regulation	Knockout	Resistance against pathogen of bacteria and fungi	Thomazella *et al.* (2016)
Pseudomonas syringae pv.	Bacterial speck	*S. lycopersicum*	*SlJAZ-2*	*S* gene of host	Produced coronatine	Knockout	Increased resistance to bacterial speck disease	Ortigosa *et al.* (2019)
Erwinia amylovora	Fire blight	*Golden delicious*	*DIPM*1 *DIPM*2 *DIPM*4	S gene of host	Interaction of pathogen effectors	Knockout	Increase resistance by using RNP method	Malnoy *et al.* (2016)
Xanthomonas citri	Citrus canker disease	Citrus species	*CsLOB*-1	S gene of host	Assistance in Infection of pathogen	Knockout	Resistance against canker disease	Jia *et al.* (2016), Peng *et al.* (2017)
Burkholderia glumae	Bacterial blight	*Oryza sativa*	*OsMPK*-5	Host S gene	Assist pathogen infection	Knockout	CRISPR editing increase resistance	Xie and Yang (2013)
Xanthomonas oryzae	Bacterial blight	*O. sativa*	*OsSWEET*11, *OsSWEET*14	*S* gene of host	Transportation of sucrose	Knockout	Editing susceptibility gene	Jiang *et al.* (2013)

ACTIVATION OF DEFENSE GENES THROUGH CRISPR ACTIVATOR (CRISPRa)

By assisting in the creation of cutting-edge technologies that overcome some of the drawbacks of conventional genetic editing, CRISPR-Cas9 has transformed a number of fields in plant science. When it comes to creating plants with outstanding agronomic qualities, the discovery of inducible CRISPR-Cas9 transcriptional activator techniques (CRISPRa) holds out a lot of potential. In order to increase gene expression, CRISPRa is a sort of CRISPR tool that combines transcriptional activators with a modified Cas9 variant lacking endonuclease function (dead Cas protein; dCas). CRISPR-dCas9 eliminates the endonuclease cleavage activity but keeps the ability to bind the targeted DNA sequence when a deactivated form of the Cas9 protein is made by tweaking its nuclease domains (Chen & Qi, 2017). Any gene may be precisely and successfully activated without inducing any alterations into the endogenous gene by fusing dCas9 with activation domains. The most popular kind of such CRISPRa activator is *Streptococcus pyogenes* (SP)-dCas9 coupled with VP64 (and combinations) transcriptional activator domains, which has been shown to increase endogenous expression.

A well-known transcription activator for the herpes simplex virus, VP64 is a tetramer of VP16 (Di Maria *et al.* 2020). It belongs to the first generation of CRISPRa systems, and several investigations show a significant induction of activation. The *dCas9* gene editing tool is combined with transactivator and sgRNA in first-generation CRISPRa systems (Shakirova *et al.* 2020). Gene activation also uses other first-generation dCas transactivators, including p65 and p300. As activators, various oligomers of VP16, VP48, VP160 or VP192 have been used. According to research, the deactivated CRISPR-dCas9 system may be combined with four tandem repetitions of the transcriptional activator VP16 or VP64 to activate the endogenous genes in those plants that were affected by bacterial pathogens. In Arabidopsis and tobacco, it was reported that protein transcriptional activation had been effective (Li *et al.* 2017). Furthermore, the low to moderate activation rates in plants were only seen in this first generation of the CRISPRa system with single domain fusions to dCas9. Consequently, the CRISPRa system of the second generation was created.

Now we utilized CRISPRa to enhance the expression of endogenous banana genes, such as disease resistance *R* genes, pathogenesis-related proteins (PRP), antibacterial Vicilin, leucine-rich repeat (LRR) and wall associated kinase (Wak2 and Wak5). Based on the transcriptome investigation of the BXW-resistant diploid banana progenitor '*Musa balbisiana*' and the BXW-susceptible cultivar 'Pisang Awak', the genes were discovered (Tripathi *et al.* 2021).

Because the dCas9 fusion protein must be continuously expressed when using the CRISPRa method, it will always be transgenic and subject to biosafety rules. Other methods, such as modifying promoters to activate endogenous genes should be investigated in order to boost expression in a nontransgenic way.

CONCLUSION AND FUTURE PROSPECTS

The findings covered at this point indicate how CRISPR-Cas system can be used to create confrontation against a wide range of bacteria. The CRISPR-Cas system is multifaceted; it has broadened our horizons in the field of genetic engineering and allowed us to discover the extraordinary and intricate molecular secrets that exist within biological systems. Although, there are issues that must be addressed. However, utilizing GE by CRISPR to build resistance to plant diseases is a potential way to overcome breeding constraints. In plant–pathogen interaction investigations, several 'Omics' methods, for example, metabolomics, transcriptomics and proteomics are required to study defense mechanisms in plants (Patterson *et al.* 2019). The findings of these investigations may lead to the identification of a novel type of cellular targets that can aid in the development of more resistant plant cultivars.

In the future, it is anticipated that the CRISPR-Cas system will be utilizing in conjunction with other approaches to develop disease-resistant plants that can tolerate biotic and environmental pressures while producing an adequate amount of food for the growing population. Some edited crops that have been created with the help of CRISPR are still awaiting regulatory approval. They may gain acceptance by developing new crops that may consist of more desirable features such as therapeutic capabilities (in the form of an edible vaccine) and increased yield.

Every scientific discipline is taking advantage of CRISPR-Cas technology, providing sustainable agriculture as a potent and adaptable gene editing and control tool. In the future, we predict immense use of CRISPR technology will significantly increase our understanding in disease resistance as well as fundamental biology.

REFERENCES

Ahmad, S., Wei, X., Sheng, Z., Hu, P., & Tang, S. (2020). CRISPR/Cas9 for development of disease resistance in plants: Recent progress, limitations and future prospects. *Briefings in functional genomics, 19*(1), 26–39.

Al-Saadi, A., Reddy, J. D., Duan, Y. P., Brunings, A. M., Yuan, Q., & Gabriel, D. W. (2007). All five host-range variants of Xanthomonas citri carry one pthA homolog with 17.5 repeats that determines pathogenicity on citrus, but none determine host-range variation. *Molecular Plant-Microbe Interactions, 20*(8), 934–943.

Aman, R., Ali, Z., Butt, H., Mahas, A., Aljedaani, F., Khan, M. Z., . . . Mahfouz, M. (2018). RNA virus interference via CRISPR/Cas13a system in plants. *Genome Biology, 19*(1), 1–9.

Andersen, E. J., Ali, S., Byamukama, E., Yen, Y., & Nepal, M. P. (2018). Disease resistance mechanisms in plants. *Genes, 9*(7), 339.

Bhaya, D., Davison, M., & Barrangou, R. (2011). CRISPR-Cas systems in bacteria and archaea: Versatile small RNAs for adaptive defense and regulation. *Annual Review of Genetics, 45*, 273–297.

Borrelli, V. M., Brambilla, V., Rogowsky, P., Marocco, A., & Lanubile, A. (2018). The enhancement of plant disease resistance using CRISPR/Cas9 technology. *Frontiers in Plant Science*, 1245.

Brooks, C., Nekrasov, V., Lippman, Z. B., & Van Eck, J. (2014). Efficient gene editing in tomato in the first generation using the clustered regularly interspaced short palindromic repeats/CRISPR-associated9 system. *Plant Physiology, 166*(3), 1292–1297.

Cernadas, R. A., Doyle, E. L., Niño-Liu, D. O., Wilkins, K. E., Bancroft, T., Wang, L., . . . White, F. F. (2014). Code-assisted discovery of TAL effector targets in bacterial leaf streak of rice reveals contrast with bacterial blight and a novel susceptibility gene. *PLoS Pathogens, 10*(2), e1003972.

Chen, K., Wang, Y., Zhang, R., Zhang, H., & Gao, C. (2019). CRISPR/Cas genome editing and precision plant breeding in agriculture. *Annual Review of Plant Biology, 70*, 667–697.

Chen, M., & Qi, L. S. (2017). Repurposing CRISPR system for transcriptional activation. *RNA Activation*, 147–157.

Chu, Z., Yuan, M., Yao, J., Ge, X., Yuan, B., Xu, C., . . . Bennetzen, J. L. (2006). Promoter mutations of an essential gene for pollen development result in disease resistance in rice. *Genes & Development, 20*(10), 1250–1255.

Das, A., Sharma, N., & Prasad, M. (2019). CRISPR/Cas9: a novel weapon in the arsenal to combat plant diseases. *Frontiers in Plant Science, 9*, 2008.

Di Maria, V., Moindrot, M., Ryde, M., Bono, A., Quintino, L., & Ledri, M. (2020). Development and validation of CRISPR activator systems for overexpression of CB1 receptors in neurons. *Frontiers in Molecular Neuroscience, 13*, 168.

Dominguez, A. A., Lim, W. A., & Qi, L. S. (2016). Beyond editing: Repurposing CRISPR–Cas9 for precision genome regulation and interrogation. *Nature Reviews Molecular Cell Biology, 17*(1), 5–15.

Dong, O. X., & Ronald, P. C. (2019). Genetic engineering for disease resistance in plants: Recent progress and future perspectives. *Plant Physiology, 180*(1), 26–38.

Fauser, F., Schiml, S., & Puchta, H. (2014). Both CRISPR/C as-based nucleases and nickases can be used efficiently for genome engineering in A rabidopsis thaliana. *The Plant Journal, 79*(2), 348–359.

Gimenez-Ibanez, S., Boter, M., Ortigosa, A., García-Casado, G., Chini, A., Lewsey, M. G., . . . Solano, R. (2017). JAZ 2 controls stomata dynamics during bacterial invasion. *New Phytologist, 213*(3), 1378–1392.

Hu, N., Xian, Z., Li, N., Liu, Y., Huang, W., Yan, F., Li, Z. (2019). Rapid and user-friendly open-source CRISPR/Cas9 system for single-or multi-site editing of tomato genome. *Horticulture research, 6*.

Hu, Y., Zhang, J., Jia, H., Sosso, D., Li, T., Frommer, W. B., . . . Jones, J. B. (2014). Lateral organ boundaries 1 is a disease susceptibility gene for citrus bacterial canker disease. *Proceedings of the National Academy of Sciences, 111*(4), E521–E529.

Jia, H., Orbovic, V., Jones, J. B., & Wang, N. (2016). Modification of the PthA4 effector binding elements in Type I Cs LOB 1 promoter using Cas9/sg RNA to produce transgenic Duncan grapefruit alleviating XccΔpthA4: dCs LOB 1.3 infection. *Plant Biotechnology Journal, 14*(5), 1291–1301.

Jia, H., Wang, Y., Su, H., Huang, X., & Wang, N. (2022). LbCas12a-D156R efficiently edits LOB1 effector binding elements to generate canker-resistant citrus plants. *Cells, 11*(3), 315.

Jiang, W., Zhou, H., Bi, H., Fromm, M., Yang, B., & Weeks, D. P. (2013). Demonstration of CRISPR/Cas9/sgRNA-mediated targeted gene modification in Arabidopsis, tobacco, sorghum and rice. *Nucleic Acids Research, 41*(20), e188–e188.

Jinek, M., Chylinski, K., Fonfara, I., Hauer, M., Doudna, J. A., & Charpentier, E. (2012). A programmable dual-RNA–guided DNA endonuclease in adaptive bacterial immunity. *Science, 337*(6096), 816–821.

Kachroo, A., Vincelli, P., & Kachroo, P. (2017). Signaling mechanisms underlying resistance responses: What have we learned, and how is it being applied? *Phytopathology, 107*(12), 1452–1461.

Kamber, T., Pothier, J. F., Pelludat, C., Rezzonico, F., Duffy, B., & Smits, T. H. (2017). Role of the type VI secretion systems during disease interactions of Erwinia amylovora with its plant host. *BMC Genomics, 18*(1), 1–12.

Kamoun, S. (2006). A catalogue of the effector secretome of plant pathogenic oomycetes. *Annual Review of Phytopathology, 44*, 41–60.

Kay, S., Hahn, S., Marois, E., Wieduwild, R., & Bonas, U. (2009). Detailed analysis of the DNA recognition motifs of the Xanthomonas type III effectors AvrBs3 and AvrBs3Δrep16. *The Plant Journal, 59*(6), 859–871.

Kubicek, C. P., Starr, T. L., & Glass, N. L. (2014). Plant cell wall–degrading enzymes and their secretion in plant-pathogenic fungi. *Annual Review of Phytopathology, 52*, 427–451.

Kumar, M., Jaiswal, S., Sodhi, K. K., Shree, P., Singh, D. K., Agrawal, P. K., & Shukla, P. (2019). Antibiotics bioremediation: Perspectives on its ecotoxicity and resistance. *Environment International, 124*, 448–461.

Kumar Singh, P., & Shukla, P. (2015). Systems biology as an approach for deciphering microbial interactions. *Briefings in Functional Genomics, 14*(2), 166–168.

Langner, T., Kamoun, S., & Belhaj, K. (2018). CRISPR crops: Plant genome editing toward disease resistance. *Annual Review of Phytopathology, 56*, 479–512.

Li, Z., Zhang, D., Xiong, X., Yan, B., Xie, W., Sheen, J., & Li, J.-F. (2017). A potent Cas9-derived gene activator for plant and mammalian cells. *Nature Plants, 3*(12), 930–936.

Liu, D., Hu, R., Palla, K. J., Tuskan, G. A., & Yang, X. (2016). Advances and perspectives on the use of CRISPR/Cas9 systems in plant genomics research. *Current Opinion in Plant Biology, 30*, 70–77.

Liu, M., Rehman, S., Tang, X., Gu, K., Fan, Q., Chen, D., & Ma, W. (2019). Methodologies for improving HDR efficiency. *Frontiers in Genetics, 9*, 691.

Lone, B. A., Karna, S. K. L., Ahmad, F., Shahi, N., & Pokharel, Y. R. (2018). CRISPR/Cas9 system: A bacterial tailor for genomic engineering. *Genetics Research International, 2018*.

Luu, D. D., Joe, A., Chen, Y., Parys, K., Bahar, O., Pruitt, R., . . . Adamchak, C. (2019). Biosynthesis and secretion of the microbial sulfated peptide RaxX and binding to the rice XA21 immune receptor. *Proceedings of the National Academy of Sciences, 116*(17), 8525–8534.

Makarova, K. S., Wolf, Y. I., Alkhnbashi, O. S., Costa, F., Shah, S. A., Saunders, S. J., . . . Haft, D. H. (2015). An updated evolutionary classification of CRISPR–Cas systems. *Nature Reviews Microbiology, 13*(11), 722–736.

Malnoy, M., Viola, R., Jung, M.-H., Koo, O.-J., Kim, S., Kim, J.-S., . . . Nagamangala Kanchiswamy, C. (2016). DNA-free genetically edited grapevine and apple protoplast using CRISPR/Cas9 ribonucleoproteins. *Frontiers in Plant Science, 7*, 1904.

Modrzejewski, D., Hartung, F., Sprink, T., Krause, D., Kohl, C., Schiemann, J., & Wilhelm, R. (2018). What is the available evidence for the application of genome editing as a new tool for plant trait modification and the potential occurrence of associated off-target effects: A systematic map protocol. *Environmental Evidence, 7*(1), 1–8.

Moscou, M. J., & Bogdanove, A. J. (2009). A simple cipher governs DNA recognition by TAL effectors. *Science, 326*(5959), 1501–1501.

Mushtaq, M., Sakina, A., Wani, S. H., Shikari, A. B., Tripathi, P., Zaid, A., . . . Singh, A. K. (2019). Harnessing genome editing techniques to engineer disease resistance in plants. *Frontiers in Plant Science, 10*, 550.

Ortigosa, A., Gimenez-Ibanez, S., Leonhardt, N., & Solano, R. (2019). Design of a bacterial speck resistant tomato by CRISPR/Cas9-mediated editing of Sl JAZ 2. *Plant Biotechnology Journal, 17*(3), 665–673.

Oskar, M., Buchwald, W., & Nawrot, R. (2015). Plant defense responses against viral and bacterial pathogen infections. Focus on RNA-binding proteins (RBPs). *Herba Polonica, 60*(4), 60–73.

Patterson, E. L., Saski, C., Küpper, A., Beffa, R., & Gaines, T. A. (2019). Omics potential in herbicide-resistant weed management. *Plants, 8*(12), 607.

Peng, A., Chen, S., Lei, T., Xu, L., He, Y., Wu, L., . . . Zou, X. (2017). Engineering canker-resistant plants through CRISPR/Cas9-targeted editing of the susceptibility gene Cs LOB 1 promoter in citrus. *Plant Biotechnology Journal, 15*(12), 1509–1519.

Pickar-Oliver, A., & Gersbach, C. A. (2019). The next generation of CRISPR–Cas technologies and applications. *Nature Reviews Molecular Cell biology, 20*(8), 490–507.

Pritchard, L., & Birch, P. R. (2014). The zigzag model of plant–microbe interactions: Is it time to move on? *Molecular Plant Pathology, 15*(9), 865.

Ricroch, A., Clairand, P., & Harwood, W. (2017). Use of CRISPR systems in plant genome editing: Toward new opportunities in agriculture. *Emerging Topics in Life Sciences, 1*(2), 169–182.

Rosa, E., Woestmann, L., Biere, A., & Saastamoinen, M. (2018). A plant pathogen modulates the effects of secondary metabolites on the performance and immune function of an insect herbivore. *Oikos, 127*(10), 1539–1549.

Schwartz, A. R., Potnis, N., Timilsina, S., Wilson, M., Patané, J., Martins Jr, J., . . . Almeida, N. (2015). Phylogenomics of Xanthomonas field strains infecting pepper and tomato reveals diversity in effector repertoires and identifies determinants of host specificity. *Frontiers in Microbiology, 6*, 535.

Shakirova, K. M., Ovchinnikova, V. Y., & Dashinimaev, E. B. (2020). Cell reprogramming with CRISPR/Cas9 based transcriptional regulation systems. *Frontiers in Bioengineering and Biotechnology, 8*, 882.

Silva, M. S., Arraes, F. B. M., de Araújo Campos, M., Grossi-de-Sa, M., Fernandez, D., de Souza Cândido, E., . . . Grossi-de-Sa, M. F. (2018). Potential biotechnological assets related to plant immunity modulation applicable in engineering disease-resistant crops. *Plant Science, 270*, 72–84.

Sugio, A., Yang, B., Zhu, T., & White, F. F. (2007). Two type III effector genes of Xanthomonas oryzae pv. oryzae control the induction of the host genes OsTFIIAγ1 and OsTFX1 during bacterial blight of rice. *Proceedings of the National Academy of Sciences, 104*(25), 10720–10725.

Tripathi, L., Ntui, V. O., Tripathi, J. N., & Kumar, P. L. (2021). Application of CRISPR/Cas for diagnosis and management of viral diseases of banana. *Frontiers in Microbiology, 11*, 609784.

Tyagi, S., Mulla, S. I., Lee, K.-J., Chae, J.-C., & Shukla, P. (2018). VOCs-mediated hormonal signaling and crosstalk with plant growth promoting microbes. *Critical Reviews in Biotechnology, 38*(8), 1277–1296.

Vale, F. X. R., Parlevliet, J., & Zambolim, L. (2001). Concepts in plant disease resistance. *Fitopatologia Brasileira, 26*(3), 577–589.

Xie, K., & Yang, Y. (2013). RNA-guided genome editing in plants using a CRISPR–Cas system. *Molecular Plant, 6*(6), 1975–1983.

Yang, B., & White, F. F. (2004). Diverse members of the AvrBs3/PthA family of type III effectors are major virulence determinants in bacterial blight disease of rice. *Molecular Plant-Microbe Interactions, 17*(11), 1192–1200.

Yin, K., & Qiu, J.-L. (2019). Genome editing for plant disease resistance: Applications and perspectives. *Philosophical Transactions of the Royal Society B, 374*(1767), 20180322.

Yu, Y., Streubel, J., Balzergue, S., Champion, A., Boch, J., Koebnik, R., . . . Szurek, B. (2011). Colonization of rice leaf blades by an African strain of Xanthomonas oryzae pv. oryzae depends on a new TAL effector that induces the rice nodulin-3 Os11N3 gene. *Molecular Plant-Microbe Interactions, 24*(9), 1102–1113.

Zeilmaker, T., Ludwig, N. R., Elberse, J., Seidl, M. F., Berke, L., Van Doorn, A., . . . Van den Ackerveken, G. (2015). DOWNY MILDEW RESISTANT 6 and DMR 6-LIKE OXYGENASE 1 are partially redundant but distinct suppressors of immunity in Arabidopsis. *The Plant Journal, 81*(2), 210–222.

Zeng, W., Melotto, M., & He, S. Y. (2010). Plant stomata: A checkpoint of host immunity and pathogen virulence. *Current Opinion in Biotechnology, 21*(5), 599–603.

Zhou, H., Liu, B., Weeks, D. P., Spalding, M. H., & Yang, B. (2014). Large chromosomal deletions and heritable small genetic changes induced by CRISPR/Cas9 in rice. *Nucleic Acids Research, 42*(17), 10903–10914.

7 CRISPR-Cas 9 System to Combat Plant Fungal Infections

INTRODUCTION

The English word 'fungus' is directly derived from the Latin word 'fungus' (Ravichandra, 2013). Fungus, plural fungi, possess approximately 80,000 known species belonging to the kingdom Fungi. Fungi include rusts, smuts, molds and mushrooms. Fungi are eukaryotic organisms having membrane-bound organelles and separate nuclei in each cell. At first, fungi were considered part of the kingdom Plantae. But soon fungi were separated from plants, because they lack chlorophyll, had different cell wall compositions and altered modes of nutrition. A typical fungus body or thallus is made up of mycelium. The fungus reproduces by generating spores directly or, in particular, fruiting bodies. Fungi can exist in air, water and soil. Fungi are unable to perform photosynthesis, and hence they collect nutrients by secreting enzymes onto the surface of organisms they are growing on. These enzymes break down organic material to produce nutrient-rich solutions that mycelium can absorb. Fungi release carbon, phosphorus, oxygen and nitrogen into the soil or atmosphere as a result of the decomposition of organic compounds. Fungi are utilized in the synthesis of enzymes, antibiotics, organic acids and vitamins which are vital for many food and industrial processes. Fungi can harm crops, infect people with diseases like candidiasis, and can cause mildew and rot in clothing and food. Some fungi have symbiotic relations with algae (forming lichens), insects, plants (forming mycorrhizae) and other living beings. Parasitic fungi attack living things frequently, bringing about infection and death.

Luckily for plants, their interactions with fungus are frequently advantageous to both the parties (endophytes, saprophytic fungi and mycorrhizae). A small percentage of fungal species has progressed and disturbed the balance of mutual benefit to cause plant diseases (Grayer & Kokubun, 2001). As a result, plants are constantly subject to biotic stress. The majority of infections that occur in horticultural and agricultural settings are caused by plant fungal pathogens. Collectively, the phytopathogens have devised strategies and tactics to harm any plant that is actively looking to gain access to resources for growth and development. These diseases have sexual or asexual origin, and they are capable of defeating a plant's immune system. This harms the health, homeostasis, physiology and, in some situations, the systemic health of the plant.

Along with agricultural and cultural practices, pesticides can offer some degree of protection toward plant disease. However, pesticides are frequently the target of strict

DOI: 10.1201/b22901-7

control due to their ability to inflict catastrophic environmental harm and their swift loss of effectiveness as a result of mutation in pathogens (Tyagi *et al.* 2021). By developing plant varieties with inbuilt disease resistance through breeding, plant diseases can be managed in an ecofriendly and frequently complimentary approach (Sánchez-Martín & Keller, 2019). Future breeding programs should identify plants with a high level of disease resistance characteristics. Adopting modern breeding approaches will facilitate pyramiding of desired traits. Conventional breeding, transgenic technology and mutation breeding have all been successful over the past few decades, but there are some drawbacks. In recent years, to overcome these limitations, the integration of novel biotechnological techniques such as transcription activator-like (TAL) effector nucleases (TALENs), zinc finger nucleases (ZFNs), meganucleases (MNs) and clustered regularly interspaced palindromic repeats (CRISPR)-Cas endonucleases have been introduced.

FUNGAL PATHOGENS IN PLANTS

Many plant diseases are caused by fungi and have a significant economic impact on agriculture. They are responsible for a variety of diseases, including rusts, rots, smuts, and mildews. Mycotoxins, which are produced by several fungal diseases, may have devastating consequences on human health (Zaynab *et al.* 2020). Diseases caused by fungi spread rapidly under favorable weather conditions and cause extensive damage to crops. For instance, botrytis bunch rot, a fungus, may cause approximately 80% loss of strawberry fruit (Petrasch *et al. 2020*). Fungus can also attack on tomato plants, causing up to 70% of tomatoes to rot before they reach maturity. Approximately 1,400 plant species are vulnerable to infection by this fungus, including grapes, beans, raspberries, blackberries, lettuce, and flowers like roses, orchids, and dahlias. Epidemics caused by fungi may have devastating effects on people. For instance, the Great Famine of Ireland (1845–1852) was caused by a disease called potato late blight fungus, which destroyed potatoes, the main source of food for the poor at the time. This resulted in the deaths of one million people. Examples of plant diseases induced by fungus are given in Table 7.1.

FUNGAL PATHOGEN DISEASE CYCLE

Phytopathogenic fungi can grow on or inside the plants. These are autoecious and heteroecious in nature. The pathogenic fungus, particular to a plant host, effectively grows with the host plant life cycle. The fungi over-winter or over-summer by forming resting spores, a thick-walled spores which can resist the severe environmental conditions. Sclerotia are a type of hyphae that can be made by soil-borne fungi when they are in hard environments. They are made up of a dense mass of hyphae. When the weather is good, the sclerotium makes new fungal growth or structures that make spores. When conditions aren't good, some fungi hide inside the plant or move to a different host plant. Finally, some fungi live in the ground, and eat dead plant and animal matter to stay alive during the winter.

TABLE 7.1
Fungal Diseases in Plants

Sr. No.	Host Plant	Causative Fungi	Disease	Reference
1	Capsicum assamicum	*Cercospora tezpurensis* sp. *nov.*	Leaf spot	Meghvansi *et al.* (2013)
2	Wheat	*Puccinia striiformis*	Stripe rust	Wang *et al.* (2009)
3	Cannabis sativa	*Golovinomyces cichoracearum sensu lato*	Hemp powdery mildew	Pépin *et al.* (2018)
4	*Dioscorea* spp. Pomegranate	*Colletotrichum gloeosporioides, Pilidiella granati*	Greater yam anthracnose Twig blight and crown rot	Yang *et al.* (2017)
5	Maize	*Fusarium verticillioides* and *F. subglutinans*	Stalk rot and ear rot	Faria *et al.* (2011)
6	C. *sativa*	*Cercospora* cf. *flagellaris*	Hemp leaf spot	Doyle *et al.* (2019)
7	Barely	*Ramularia*	Ramularia leaf spot	Havis *et al.* (2014)
8	Strawberry	*Phytophthora cactorum*	Crown rot	Bhat and Browne (2010)
9	Rice	*Rhizoctonia solani*	Rice blast	Sun *et al.* (2015)
10	Cucurbits	*Didymella bryoniae*	Gummy stem blight	Ling *et al.* (2010)
11	Apple	*Neofabraea alba, N. perennans* and *N. keinholzii*	Bull's eye rot	Kirby (2016)
12	Tobacco	*Collo-cygni*	Target spot	Petrov *et al.* (2007)

When the host plant is ready in the spring, the fungus is ready to attack. The fungus makes spores, which can be carried to the plant by air, water or even insects. When spores land on a plant, they stick to the surface and cause an infection. Colonization and penetration of the host tissues are the two stages of an infection. The fungus must get into the tissues of the host to get to the nutrients inside. Fungus may enter via a plant's natural openings like stomata, degrade cuticle by enzymatic action or penetrate through existing wounds. Certain fungi invade plants by damaging their surface using specific components or structures. During colonization, the penetrating hyphae develops a specialized structure called 'haustoria' which pumps up the nutrients from the cells for its vergetative and reproductive growth. Some fungi directly devour plant cells for nutrition, while others employ hyphae to acquire nutrients from living cells.

Some fungal infections only have one infection cycle each season, whereas others have multiple by disseminating new spores after colonization. If the weather is excellent, several secondary infection cycles develop, hurting plants and crops (Agrios, 2005).

COMMON SYMPTOMS AND SIGNS OF FUNGAL DISEASES

Symptoms are the outward manifestations of a disease caused by a pathogen. Because different fungi are able to infect different plants and plant components, fungal infections may produce a broad range of symptoms, including changes in color and shape, decay, sores and wilting. The yellowing of leaves is one indication of a larger pattern of color changes that might be followed by noticeable morphological abnormalities. Spots on leaves and rotting spots on fruits are both signs of cell death, which leads to the breakdown of certain plant components and the darkening of plant tissues. The mycelium and spores of fungi, which are visible on the surface of infected plants, are not always indicators of a disease. Several fungal infections cause small, scab-like sores on plant parts. We need to know what kind of fungus is causing the plant's disease so that we can take the right preventative measures. The signs and symptoms of the disease are used for this purpose.

PLANTS NATURAL DEFENSE SYSTEM AGAINST FUNGAL PATHOGENS

Plants have ability to recognize pathogens as 'non-self'. To trigger defense mechanisms, a wide variety of signals from fungi and their surroundings are recognized by plants (Brown & Ogle, 1997). Plants use a combination of strategies from two different mechanisms to defend themselves against pathogens, known as host resistance:

1. Physical barriers act as obstructions to help in preventing pathogen entry and spread in the plants internal systems.
2. In the tissues and cells of the plant, biochemical processes produce substances that are either harmful to pathogens or that create an environment that prevents their growth. Different host–pathogen systems are made up of structural characteristics and metabolic mechanisms that are employed by plants to defend themselves.

Physical Barrier

Adhesion to cell surfaces is critical to the development of many different diseases. Examples of the subsequent production of cutin-degrading enzymes during adhesion and penetration support the conclusion that the cuticle is essential as a protective barrier. While many saprophytic fungi and bacteria do produce cutin-degrading enzymes, their primary role is to facilitate the digestion of cellulose contained in plant cell walls. Invading microbes use cutinase and other enzymes to breach the cell membrane. The cuticle at the site of penetration acts as a barrier against the invading pathogens secrete enzymes to degrade it. *Fusarrum solani* and *Fusarrum* sp. *pisi* isolates from diseased pea stem release a cutinolytic enzyme (Brown & Ogle, 1997). Leaves and fruits with wax coatings have a water-repellent surface that prevents the formation of a water film that may serve as a breeding ground for pathogenic fungi. To stave against fungal diseases, plants rely on traits like thick and tough layer of

TABLE 7.2
Antifungal Compounds to Inhibit Fungal Growth

Sr. No.	Fungus	Antifungal Compounds	References
1	*Botrytis cinerea*	Benzaldehyde Ethyl benzoate	Wilson and Wisniewski (1989)
2	*Colletotrichum circinans*	Catechol Protocatcchuic acid	Walker and Stahmann (1955)
3	*Botrytis cinerea* *Rhizopus stolonifera*	2,5-Dimethoxybenzoic acid	Lattanzio et al. (1996)
4	*Eutypa Lata*	Salicylic acid	Amborabé et al. (2002)
5	*Phytophthora infestans*	Vanillic acid 4-Hydroxybenzoic acid	Harborne (1980)
6	*Phytophthora infestans,* *Phlyctaena vagabunda*	Chlorogenic acid, Rutin	Carrasco et al. (1978)
7	*Phytophthora* spp.	Oleuropein	Del Río et al. (2003)

epidermal cells. Many dangerous fungi can only infect plants via their stomata. The structure of certain stomata, such as massive, elevated guard cells and very small openings, makes them resistant to some fungal diseases (Jibril *et al.* 2016). Cell wall proteins and enzymes actively alter wall form throughout cell growth, increasing wall thickness and strength in response to an external threat. In reaction to microbial invasion, plant cells also manufacture cellulose and deposit it between the cell wall and cell membrane. Polysaccharide polymers called calcareous deposits or papillae form at the site of infection as part of the activated innate immune response. The main function in defense against fungal infections could not be played by the structural or cellular defense.

Biochemical Defense

Plants may resist infection in some cases by not supplying the pathogen with the necessary nutrients. The germination of dormant spores of pathogens like *Spongospora subterranea* (powdery scab of potatoes), *Plasmodtophora brassicae* (club root of crucifers) and *Urocgstls agropgri* (flag or leaf smut of wheat) depends on the presence of particular chemicals. These are found in various plants' secretions, including those of potential hosts. By default, plants that don't release these stimulators are resistant (Brown & Ogle, 1997). Plants frequently contain preformed antibiotic substances such as phenolic and polyphenolic chemicals, which are essential for nonhost resistance to filamentous fungus. Some antifungal phenolics are given in Table 7.2.

General mechanism of fungal infection and plant resistance mechanism is described in Figure 7.1.

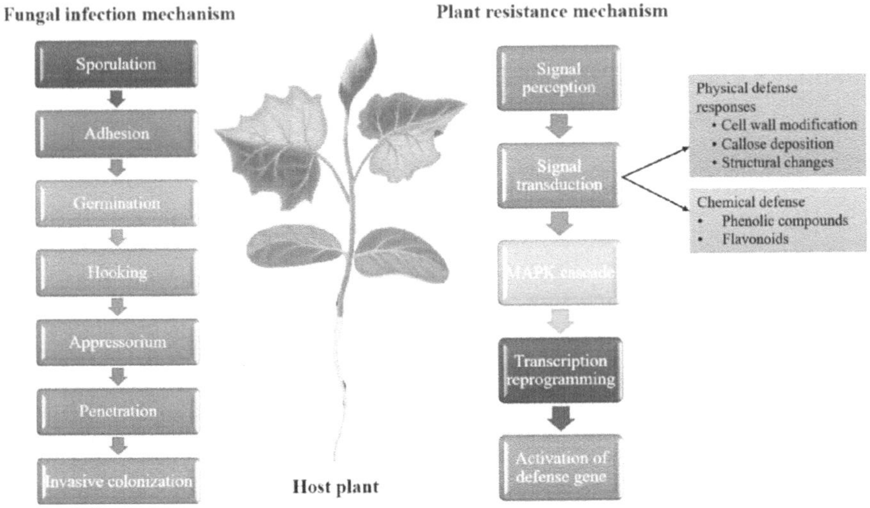

FIGURE 7.1 Fungal infection and plant resistance mechanism.

GENOME-EDITING (GE) TECHNOLOGY

In genome-editing approaches, sequence-specific nucleases (SSN) are used to detect and target certain DNA sequences and insert double-strand break (DSB) at particular target positions. Numerous mechanisms, including microhomology-mediated end-joining (MMEJ), homology-directed repair (HDR) and non-homologous end-joining (NHEJ) are employed to restore DSBs (Molla & Yang, 2020). The basic mechanism for restoring DSBs is NHEJ, although this method frequently results in errors and insertion or omission changes at the DSB sites. On the other hand, whereas it happens seldom in plant systems, the HDR route is error-free in repairing DSBs. Currently, the CRISPR-Cas system, TALENs and ZFNs are the three primary SSN types employed for genome editing (Karmakar *et al.* 2020). TALENs and ZFNs perform their genome-editing activity through DNA–protein interaction. Jinek et al. (2012a) subjugated the CRISPR-Cas9 system by designing small guide RNA to induce DSB in the target site. Subsequently, the CRISPR-Cas9 system demonstrates its role by DNA–RNA interaction, in contrast to ZFN and TALEN. Plant biotechnologists are very interested in using the CRISPR-Cas9 system for editing plant genome because of its affordability, easier method and increased efficacy (Sato *et al.* 2012). Random mutations are inserted into plants genome by frequently applied chemical mutagens and physical agents. However, specific chromosomal sites may get altered as a result of these insertion. Therefore, compared to random mutagenesis, genome-editing technology is greatly precise and offers a wider range of uses. In order to introduce mutations at marked genomic regions of various plants, the CRISPR-Cas9 system is currently widely used (Mushtaq and Molla, 2021). Several researchers have

TABLE 7.3
Genome Editing in Different Plants against Fungal Pathogens

Sr. No.	Fungal Pathogen	Disease	Host Plant	Target Gene	Reference
1	*Magnaporthe oryzae*	Rice blast	Rice	*Os*BSR-K	Zhou *et al.* (2018)
2	*Fusarium graminearum, Blumeria graminis*	Powdery mildew	Barley	*Hv*MORC1	Kumar *et al.* (2018)
3	*Fusarium graminearum*	Fusarium head blight	Wheat	*Ta*NFXL1	Brauer *et al.* (2020)
4	*Powdery mildew fungus*	Powdery mildew	Wheat	*Ta*MLO-B1	Brauer *et al.* (2020)
5	*Erysiphe necator*	Powdery mildew	Grape	*Vv*Mlo3	Guo *et al.* (2019)
6	*Sclerotinia sclerotiorum*	Sclerotinia stem rot	Rapeseed	*Bn*WRKY70 and *Bn*WRKY11	Sun *et al.* (2018)

applied methods based on ZFN and TALEN to perform clear-cut and accurate plant genome editing. Genome editing against various fungal diseases in different plants is described in Table 7.3.

ZFNs

In eukaryotes, zinc finger nucleases (ZFNs) are considered as the most primitive sequence-specific nucleases employed for genome editing. In 1985, zinc finger protein, which is a prevalent DNA binding domain in eukaryotes, was foremost recognized as a component of transcription factor IIIa in the *Xenopus oocytes*. A variety of Cys2His2 zinc fingers make up the zinc finger (ZF) domain. About 30 amino acids constitute every ZF, which binds to homologous triplets of nucleotides. With the employment of various combinations of ZF moieties, ZFs can be constructed to identify any 3–6 nucleotide triplets. There are two different functional domains in a ZFN monomer. The N-terminal area contains a synthetic ZF domain that binds to target DNA. Double-strand breaks (DSBs) are produced in concerned genomic region by FokI DNA cleavage domain, which is present in the C terminal region. More than ten years, ZFNs have been frequently employed by researchers for the concern of genome editing in both animals and plants organisms (Mushtaq and Molla, 2021). ZFN has been employed for targeted specific mutagenesis-like sequence alteration of endogenous gene ABA-INSENSITIVE-4 (*ABI4*) in Arabidopsis (Osakabe *et al.* 2010). Petolino *et al.* (2010) testified the deletion phenomena of around 4.3 kb integrated GUS gene sequence in tobacco, flanked by ZFN cleavage sites. In hexaploid wheat (*Triticum aestivum*), Ran *et al.* (2018) revealed editing of acetohydroxy acid synthase (*AHAS*) gene mediated by ZFN site-specific genome editing. They targeted three homologous copies of *AHAS* to accomplish synchronized multiple gene knockouts.

TALENs

Another synthetic protein-based DNA targeting nuclease being more precise and productive than ZFNs is called transcription activator-like effector nucleases (TALENs). A customizable target-specific DNA-binding domain is fused to a non-specific DNA cleavage domain (Fok1) in TALENs, just like in ZFNs. The -DNA binding domain in TALENs is TAL effector (TALE). TALE are proteins in nature. When *Xanthomonas* bacteria infect plants, they secrete TALEs (Malzahn *et al.* 2017). Though constructing TALENs on a big scale with identical repeat sequences is one of the main technological challenges scientists confront, it is still simpler than designing triplet confined ZFNs (Yulianto, 2018). By accurately altering the lipoxygenase (*Lox*3) gene, storage tolerance in rice has been improved using TALEN-based targeted mutagenesis (Ma *et al.* 2015). Seed viability and longevity are influenced by the lipoxygenase (*Lox*3) gene, which catalyzes the deoxygenation of polyunsaturated fatty acids. Similarly, in hexaploid bread wheat, TALEN-based mutagenesis has been introduced to target three alleles encoding Mildew-Resistance Locus O (MLO) protein. TALEN-based *NHEJ* mutation produced heritable broad-spectrum defense against powdery mildew in mutated wheat plants, a feature lacking in wild wheat (Wang *et al.* 2014).

CRISPRCas

Early in 1987, the bacterial gene CRISPR-clustered regularly interspersed short palindromic repeats—was first identified. Later, numerous other bacterial species were also found to contain CRISPR. According to how the *Cas* genes are organized and their structural variation, the CRISPR-Cas system can be largely separated into two groups. Class 1 CRISPR-Cas systems comprise of multiprotein effector complexes, while class 2 systems comprise on single effector complex. According to recent reports, there are two classes, six kinds, and more than 30 subtypes of CRISPR-Cas systems (Koonin & Makarova, 2022). The type II CRISPR-Cas9 system from *Streptococcus pyogenes* is a straightforward CRISPR-Cas system, despite the fact that there are a large variety of CRISPR-Cas systems in nature. It uses two short RNAs for target recognition and a single Cas9 endonuclease to break DNA double strands (Koonin & Makarova, 2022). To direct Cas9 for cleavage, CRISPR RNA (crRNA) binds with Trans-activating crRNA (tracrRNA) to create a hybrid crRNA-tracrRNA. Jinek *et al.* (2012) artificially combined a couple of tiny RNAs into a single RNA, facilitating the technique for genome editing, and this kind of RNA is called single guide RNA (sgRNA). As a result, the Cas9 nuclease and the sgRNA are the two major parts of the CRISPR-Cas genome-editing tool. By Watson–Crick base pairing, a Cas9-sgRNA complex binds to a 20 bp DNA target (also known as protospacer). Cas9 needs the existence of a neighboring protospacer motif (PAM) at the 3′ end of the protospacer or the target sequence.

The PAM sequence for SpCas9 is designated as 5′-NGG-3′, where N is any nucleotide. Depending on the bacterial species from which the Cas9 gene was derived, this PAM sequence varies. With the use of contemporary bioinformatics tools, guide RNAs can be constructed to produce cuts at any target DNA sequence after choosing an appropriate Cas9. For Cas9 to function, the target sequence for guide RNA must be available just after a PAM sequence. By matching of the complimentary sequence at the 5′ end of the sgRNA, the target sequence can be altered. When the Cas9-sgRNA complex detects a perfect match in the genome, it makes it easier for Cas9 to create a double-strand break (DSB) mediated by ZFNs and TALENs (Karmakar *et al.* 2020). Repair processes mistakenly introduce InDels (insertion and/or deletion) after the DSB is established, leading to frameshift mutation of the gene. Prokaryotic Cas proteins have been found in abundance, and numerous attempts at protein engineering have led to the creation of *Cas* variants that enable us to target various genomic areas (Molla *et al.* 2020).

Firstly, the guide RNAs (20 bp target sequence to be added at the 5′ end of the sgRNA) are created manually or with the aid of online resources for the purpose of executing plant genome editing. According on the kind of PAM sequences present in the target genomic region, an appropriate *Cas* variant is then selected in the second step (Molla *et al.* 2020). Third, guide RNAs' proficiency can be tested in vitro (Karmakar *et al.* 2021) or by utilizing a transient protoplast transfection system, and this phase is voluntary and lessens catastrophe. Fourth, a binary vector containing a *Cas9* expression cassette and a single or multiple sgRNA expression cassette are prepared. The binary vector is employed for transformation that is mediated by an agrobacterium. For biolistic transformation, a smaller vector containing the two cassettes or RNPs (Cas9 protein + sgRNA) can be employed as an alternative. Fifth, agrobacterium or biolistic techniques are applied to alter genetic material. Sixth, appropriate transformed cell selection and putative genome-edited plant regeneration are carried out.

Seventh, the putative plants are genotyped utilizing various techniques to identify the target region where effective genome editing occurred. After successful genome editing has been detected, transgene-free genome-edited plants can be developed by performing sexual segregation among the successful genetically modified plants. With the development of bioinformatics-based designing tools, the traditional CRISPR-mediated gene manipulation tool has made significant progress toward desired genome and protein engineering. And sparked the innovation of new applications like genome modification, transcriptional control, base editing, epigenetic modification, and prime editing (Molla *et al.* 2020).

UNDERSTANDING HOSTPATHOGEN INTERACTIONS BY GENOME EDITING

To combat infections, plants display a wide range of defense mechanisms. Pathogen-associated molecular patterns (PAMPs) or damage-associated molecular patterns are the signals that are transmitted to the cellular level when a plant comes into contact with a pathogen (DAMPs). Plants' innate immune system has two components related to each other: immunity modulation component (IMC) and immunity activation component (IAC). IAC functions through synchronized participation of receptors: -binding leucine-rich repeat (NB-LRR) receptor and pattern-recognition receptors (PRRs).

These receptors receive message of pathogen during attack by detecting appearance of infections. In contrast, the IMC immune system of plant is regulated by efficient contribution of phytohormones (Andolfo *et al.* 2016). Availability of excessive genome-editing tools has significantly accelerated our understanding regarding host-pathogen relation. Genome-editing techniques can be used to alter susceptible host genes that are responsive to pathogens or pathogen genes essential for pathogenesis and determine their precise role in infection. For example, five susceptible extracellular cystatin-like cysteine protease inhibitors (PpalEPICs) have been recognized in *Phytophthora palmivora*. Significant degree of loss of virulence was observed in *P. palmivora* against *Ppal*EPIC8 gene by using CRISPR-Cas9-based editing (Gumtow *et al.* 2018). Similarly, CRISPR-Cas12a-based editing in *Phytophthora infestans* revealed the monoallelic expression of an elicitor INF1. Fol-milR1 is a pathogenicity factor of *Fusarium oxysporum*; CRISPR-mediated knockout of Fol-milR1 exhibited decreased pathogenicity in a vulnerable tomato cultivar (Wu *et al.* 2021). Remarkably, the alike study found that the tomato *FRG*4 gene plays an important part in the resistance to tomato wilt infection since it demonstrates increased disease susceptibility when it is knocked out through CRISPR-Cas9.

USE OF GENOME-EDITING TECHNOLOGY FOR FUNGAL DISEASE RESISTANCE

The CRISPR-Cas9 method has demonstrated encouraging consequences in establishing substantial fungal disease resistance along with providing bacterial and viral resistance. Ethylene responsive factors (ERF), a subgroup of the superfamily of transcription factors known as (AP2/ERF), are involved in plant responses to both biotic and abiotic stressors. *M agnaporthe oryzae* infection causes rice to express *Os*ERF922 gene. In order to produce plants with significantly stronger resistance against *M. oryzae*, CRISPR-Cas9 technology was employed to simultaneously insert mutations at two or three sites of *Os*ERF922 (Wang *et al.* 2016). Wang et al. (2014) used the TALEN and CRISPR-Cas9 technologies on hexaploid wheat against powdery mildew disease.

These technologies allow for the introduction of heritable mutations in the MildewResistance Locus O (*MLO*) genes. Nekrasov *et al.* (2017) exploited CRISPR-Cas9 technology to create a tomato with resistance to the powdery mildew disease. "Tomelo" powdery mildew-resistant tomato cultivar was developed and claimed as non-transgenic, i.e., it is free of imported DNA sequences. As a result, this method makes it simple to introduce mutations into local or superior varieties in less than a year. One of the most significant crops is the potato, but the late blight disease severely hampers its production. In order to develop a cultivar of potatoes that is resistant to late blight, Kieu et al. (2021) screened putative *S* genes in potatoes.

The roles of these genes were predicted using bioinformatics methods, and the genes of interest were chosen. Pathogenesis is significantly assisted by susceptibility genes (*S* genes). Previous research has demonstrated that the suppression of *S* gene function increased pathogen resistance in rice, wheat, citrus and tomatoes. Tetra-allelic mutations induced by genome editing were used to create plants resistant to late blight disease (Kieu *et al.* 2021).

CRISPR-CAS9 LIMITATIONS AND FUTURE PROSPECTS

The increased availability of crop and fungal/oomycete genome sequences, as well as user-friendly CRISPR-Cas9 technologies with open resources (e.g., procedures and plasmids), has expedited disease resistance research in a variety of crops. However, there are a few constraints that must be handled. Crop plants, for example, have a somewhat extended life cycle compared to model plants, making precision and frequency of mutant production a key concern. Recently, the CRISPR-Cas9 technology was employed on cocoa using Agroinfiltration, which was researched and revealed disease resistance phenotypically (Fister *et al.* 2018). Before using CRISPR-Cas9, consider the consequences of modifying or altering plants susceptibility genes. It is well understood that vulnerable gene(s) are linked with the growth of plants and various developmental stages, resulting in pleiotropic effects (Tyagi *et al.* 2021). This is also a common occurrence after artificially increasing plant immunity by knocking off negative regulators (Ding *et al.* 2018).

Multiple analysis and testing are needed, despite the fact that numerous investigations have demonstrated that the genetic variations including deletion or mutation for these genes have no considerable impact on plant growth or health (Pyott *et al.* 2016). Furthermore, the insertion of off-target mutations by the CRISPR-Cas9 system is a crucial consideration (Ahmad *et al.* 2020). Off-target mutations are unintended consequences that induce nonspecific alterations in genomic sequences, influencing the structure and function of protein. Scientists are now striving to reduce off-target effects utilizing computational methods (Guide-seq, Diagenome-seq, DISCOVER, etc.) and re-editing CRISPR components like as gRNA and Cas proteins.

When compared to conventional molecular breeding approaches, genome editing with CRISPR-Cas9 might create desirable disease resistance within a short period of time. Due to the rapid multiplication and genetic variety of fungal populations, single gene-editing-based disease resistance may inevitably be defeated by infections. Targeting many genes or combining editing of plant and fungal genomes utilizing these rapidly evolving CRISPR-Cas9 technologies might be powerful tools for long-term disease control. The interaction of a pathogen effector and a host plant resistance gene is one idea that may be proved through combined genome editing. Editing of the *P. sojae* effector *Avr* genes (Fang & Tyler, 2016) and the soybean *Rps* genes (Nagy *et al.* 2021), can be combined and employed to control *Phytophthora* root and stem rot in soybean. This CRISPR-Cas9 gene drive mechanism might serve as a paradigm for controlling sexually transmitted plant diseases (Kyrou *et al.* 2018).

A possible method is to use the CRISPR-Cas9 system to beneficial fungal species, such as *Trichoderma* sp., to boost plant defense as a biocontrol agent against the fungal and oomycete infections. In order to increase plant disease resistance, CRISPR-Cas9-based plant and pathogen genome editing will probably be widely deployed, and disease-resistant transgene-free crops will be indispensable to meet future inclusive food demand.

REFERENCES

Agrios, G. N. (2005). *Plant Pathology*. Department of Plant Pathology University of Florida. *Elsevier* (5).

Ahmad, S., Wei, X., Sheng, Z., Hu, P., & Tang, S. (2020). CRISPR/Cas9 for development of disease resistance in plants: Recent progress, limitations and future prospects. *Briefings in Functional Genomics, 19*(1), 26–39.

Amborabé, B.-E., Fleurat-Lessard, P., Chollet, J.-F., & Roblin, G. (2002). Antifungal effects of salicylic acid and other benzoic acid derivatives towards Eutypa lata: Structure–activity relationship. *Plant Physiology and Biochemistry, 40*(12), 1051–1060.

Andolfo, G., Iovieno, P., Frusciante, L., & Ercolano, M. R. (2016). Genome-editing technologies for enhancing plant disease resistance. *Frontiers in Plant Science, 7*, 1813.

Bhat, R. G., & Browne, G. T. (2010). Specific detection of Phytophthora cactorum in diseased strawberry plants using nested polymerase chain reaction. *Plant Pathology, 59*(1), 121–129.

Brauer, E. K., Balcerzak, M., Rocheleau, H., Leung, W., Schernthaner, J., Subramaniam, R., & Ouellet, T. (2020). Genome editing of a deoxynivalenol-induced transcription factor confers resistance to Fusarium graminearum in wheat. *Molecular Plant-Microbe Interactions, 33*(3), 553–560.

Brown, J. F., & Ogle, H. J. (1997). Plant pathogens and plant diseases. Published by Rockvale Publications for the Division of Botany, University of New England.

Carrasco, A., Boudet, A. M., & Marigo, G. (1978). Enhanced resistance of tomato plants to Fusarium by controlled stimulation of their natural phenolic production. *Physiological Plant Pathology, 12*(2), 225–232.

Del Río, J. A., Báidez, A. G., Botía, J. M., & Ortuno, A. (2003). Enhancement of phenolic compounds in olive plants (Olea europaea L.) and their influence on resistance against Phytophthora sp. *Food Chemistry, 83*(1), 75–78.

Ding, Y., Sun, T., Ao, K., Peng, Y., Zhang, Y., Li, X., & Zhang, Y. (2018). Opposite roles of salicylic acid receptors NPR1 and NPR3/NPR4 in transcriptional regulation of plant immunity. *Cell, 173*(6), 1454–1467.

Doyle, V. P., Tonry, H. T., Amsden, B., Beale, J., Dixon, E., Li, H., Szarka, D., & Ward Gauthier, N. (2019). First report of Cercospora cf. Flagellaris on industrial hemp (Cannabis sativa) in Kentucky. *Plant Disease, 103*(7), 1784–1784.

Fang, Y., & Tyler, B. M. (2016). Efficient disruption and replacement of an effector gene in the oomycete P hytophthora sojae using CRISPR/C as9. *Molecular Plant Pathology, 17*(1), 127–139.

Faria, C. B., Abe, C. A. L., Silva, C. N. da, Tessmann, D. J., & Barbosa-Tessmann, I. P. (2011). New PCR assays for the identification of Fusarium verticillioides, Fusarium subglutinans, and other species of the Gibberella fujikuroi complex. *International Journal of Molecular Sciences, 13*(1), 115–132.

Fister, A. S., Landherr, L., Maximova, S. N., & Guiltinan, M. J. (2018). Transient expression of CRISPR/Cas9 machinery targeting TcNPR3 enhances defense response in Theobroma cacao. *Frontiers in Plant Science, 9*, 268.

Grayer, R. J., & Kokubun, T. (2001). Plant–fungal interactions: The search for phytoalexins and other antifungal compounds from higher plants. *Phytochemistry, 56*(3), 253–263.

Gumtow, R., Wu, D., Uchida, J., & Tian, M. (2018). A Phytophthora palmivora extracellular cystatin-like protease inhibitor targets papain to contribute to virulence on papaya. *Molecular Plant-Microbe Interactions, 31*(3), 363–373.

Guo, D.-L., Zhao, H.-L., Li, Q., Zhang, G.-H., Jiang, J.-F., Liu, C.-H., & Yu, Y.-H. (2019). Genome-wide association study of berry-related traits in grape [Vitis vinifera L.] based on genotyping-by-sequencing markers. *Horticulture Research*, 6.

Harborne, J. B. (1980). *Plant Phenolics. Secondary Plant Products*. Springer-Verlag, *717* (1), 93–118.

Havis, N. D., Gorniak, K., Carmona, M. A., Formento, A. N., Luque, A. G., & Scandiani, M. M. (2014). First molecular detection of Ramularia leaf spot (Ramularia collo-cygni) in seeds and leaves of barley in Argentina. *Plant Disease*, *98*(2), 277–277.

Jibril, S. M., Jakada, B. H., Kutama, A. S., & Umar, H. Y. (2016). Plant and pathogens: Pathogen recognision, invasion and plant defense mechanism. *International Journal of Current Microbiology and Applied Sciences*, *5*(6), 247–257.

Jinek, M., Chylinski, K., Fonfara, I., Hauer, M., Doudna, J. A., & Charpentier, E. (2012). A programmable dual-RNA–guided DNA endonuclease in adaptive bacterial immunity. *Science*, *337*(6096), 816–821.

Karmakar, S., Behera, D., Baig, M. J., & Molla, K. A. (2021). In vitro Cas9 cleavage assay to check guide RNA efficiency. In *CRISPR-Cas Methods* (pp. 23–39). Springer.

Karmakar, S., Molla, K. A., & Molla, J. (2020). Genetic Engineering and Genome Editing Strategies to Enhance Diseases Resistance of Rice Plants: A Review of Progress and Future Prospects. Rice Research for Quality Improvement: Genomics and Genetic Engineering, 35–59.

Kirby, J. N. (2016). The role of mindfulness and compassion in enhancing nurturing family environments. *Clinical Psychology: Science and Practice*, *23*(2), 142.

Koonin, E. V., & Makarova, K. S. (2022). Evolutionary plasticity and functional versatility of CRISPR systems. *PLoS Biology*, *20*(1), e3001481.

Kumar, N., Galli, M., Ordon, J., Stuttmann, J., Kogel, K.-H., & Imani, J. (2018). Further analysis of barley MORC 1 using a highly efficient RNA-guided Cas9 gene-editing system. *Plant Biotechnology Journal*, *16*(11), 1892–1903.

Kyrou, K., Hammond, A. M., Galizi, R., Kranjc, N., Burt, A., Beaghton, A. K., Nolan, T., & Crisanti, A. (2018). A CRISPR–Cas9 gene drive targeting doublesex causes complete population suppression in caged Anopheles gambiae mosquitoes. *Nature Biotechnology*, *36*(11), 1062–1066.

Lattanzio, V., Di Venere, D., Linsalata, V., Lima, G., Ippolito, A., & Salerno, M. (1996). Antifungal activity of 2, 5-dimethoxybenzoic acid on postharvest pathogens of strawberry fruits. *Postharvest Biology and Technology*, *9*(3), 325–334.

Ling, K.-S., Wechter, W. P., Somai, B. M., Walcott, R. R., & Keinath, A. P. (2010). An improved real-time PCR system for broad-spectrum detection of Didymella bryoniae, the causal agent of gummy stem blight of cucurbits. *Seed Science and Technology*, *38*(3), 692–703.

Liu, R., Chen, L., Jiang, Y., Zhou, Z., & Zou, G. (2015). Efficient genome editing in filamentous fungus Trichoderma reesei using the CRISPR/Cas9 system. *Cell Discov 1*: 15007.

Ma, L., Zhu, F., Li, Z., Zhang, J., Li, X., Dong, J., & Wang, T. (2015). TALEN-based mutagenesis of lipoxygenase LOX3 enhances the storage tolerance of rice (Oryza sativa) seeds. *PLoS One*, *10*(12), e0143877.

Malzahn, A., Lowder, L., & Qi, Y. (2017). Plant genome editing with TALEN and CRISPR. *Cell & Bioscience*, *7*(1), 1–18.

Meghvansi, M. K., Khan, M. H., Gupta, R., & Veer, V. (2013). Identification of a new species of Cercospora causing leaf spot disease in Capsicum assamicum in northeastern India. *Research in Microbiology*, *164*(9), 894–902.

Molla, K. A., Karmakar, S., & Islam, M. T. (2020). Wide horizons of CRISPR-Cas-derived technologies for basic biology, agriculture, and medicine. In *CRISPR-Cas Methods*. Springer. pp. 1–23.

Molla, K. A., & Yang, Y. (2020). Predicting CRISPR/Cas9-induced mutations for precise genome editing. *Trends in Biotechnology*, *38*(2), 136–141.

Nagy, E. D., Stevens, J. L., Yu, N., Hubmeier, C. S., LaFaver, N., Gillespie, M., Gardunia, B., Cheng, Q., Johnson, S., & Vaughn, A. L. (2021). Novel disease resistance gene paralogs created by CRISPR/Cas9 in soy. *Plant Cell Reports*, *40*(6), 1047–1058.

Nødvig, C. S., Nielsen, J. B., Kogle, M. E., & Mortensen, U. H. (2015). A CRISPR-Cas9 system for genetic engineering of filamentous fungi. *PloS One*, *10*(7), e0133085.

Osakabe, K., Osakabe, Y., & Toki, S. (2010). Site-directed mutagenesis in Arabidopsis using custom-designed zinc finger nucleases. *Proceedings of the National Academy of Sciences*, *107*(26), 12034–12039.

Petrasch, S. (2020). Genetics of Strawberry Postharvest Fruit Quality and Resistance to Necrotrophic Fungi (Doctoral dissertation, University of California, Davis). 28157405.

Pépin, N., Punja, Z. K., & Joly, D. L. (2018). Occurrence of powdery mildew caused by Golovinomyces cichoracearum sensu lato on Cannabis sativa in Canada. *Plant Disease*, *102*(12), 2644–2644.

Petolino, J. F., Worden, A., Curlee, K., Connell, J., Strange Moynahan, T. L., Larsen, C., & Russell, S. (2010). Zinc finger nuclease-mediated transgene deletion. *Plant Molecular Biology*, *73*(6), 617–628.

Petrov, R. H., García, O. L., Mulders, J.., Reis, A. C. C., Bae, J. H., Kestens, L. A., & Houbaert, Y. (2007). Three dimensional microstructure–microtexture characterization of pipeline steel. *Materials Science Forum*, *550*, 625–630.

Qin, H., Xiao, H., Zou, G., Zhou, Z., & Zhong, J. (2017). CRISPR-Cas9 assisted gene disruption in the higher fungus Ganoderma species. *Process Biochemistry*, *56*, 57–61.

Ran, Y., Patron, N., Kay, P., Wong, D., Buchanan, M., Cao, Y.-Y., Sawbridge, T., Davies, J. P., Mason, J., & Webb, S. R. (2018). Zinc finger nuclease-mediated precision genome editing of an endogenous gene in hexaploid bread wheat (Triticum aestivum) using a DNA repair template. *Plant Biotechnology Journal*, *16*(12), 2088–2101.

Ravichandra, N. G. (2013). *Fundamentals of Plant Pathology*. PHI Learning Pvt. Ltd.

Sánchez-Martín, J., & Keller, B. (2019). Contribution of recent technological advances to future resistance breeding. *Theoretical and Applied Genetics*, *132*(3), 713–732.

Sato, S., Tabata, S., Hirakawa, H., Asamizu, E., Shirasawa, K., Isobe, S., Kaneko, T., Nakamura, Y., Shibata, D., & Aoki, K. (2012). The tomato genome sequence provides insights into fleshy fruit evolution. *Nature*, *485*(7400), 635–641.

Schuster, M., Schweizer, G., Reissmann, S., & Kahmann, R. (2016). Genome editing in Ustilago maydis using the CRISPR–Cas system. *Fungal Genetics and Biology*, *89*, 3–9.

Sun, G., Liu, J., Li, G., Zhang, X., Chen, T., Chen, J., Zhang, H., Wang, D., Sun, F., & Pan, H. (2015). Quick and accurate detection and quantification of Magnaporthe oryzae in rice using real-time quantitative polymerase chain reaction. *Plant Disease*, *99*(2), 219–224.

Sun, Q., Lin, L., Liu, D., Wu, D., Fang, Y., Wu, J., & Wang, Y. (2018). CRISPR/Cas9-mediated multiplex genome editing of the BnWRKY11 and BnWRKY70 genes in Brassica napus L. *International Journal of Molecular Sciences*, *19*(9), 2716.

Tyagi, S., Kumar, R., Kumar, V., Won, S. Y., & Shukla, P. (2021). Engineering disease resistant plants through CRISPR-Cas9 technology. *GM Crops & Food*, *12*(1), 125–144.

Walker, J. C., & Stahmann, M. A. (1955). Chemical nature of disease resistance in plants. *Annual Review of Plant Physiology*, *6*(1), 351–366.

Wang, Q., Cobine, P. A., & Coleman, J. J. (2018). Efficient genome editing in Fusarium oxysporum based on CRISPR/Cas9 ribonucleoprotein complexes. *Fungal Genetics and Biology*, *117*, 21–29.

Wang, X., Tang, C., Chen, J., Buchenauer, H., Zhao, J., Han, Q., Huang, L., & Kang, Z. (2009). Detection of Puccinia striiformis in latently infected wheat leaves by nested polymerase chain reaction. *Journal of Phytopathology*, 157(7–8), 490–493.

Wang, Y., Cheng, X., Shan, Q., Zhang, Y., Liu, J., Gao, C., & Qiu, J.-L. (2014). Simultaneous editing of three homoeoalleles in hexaploid bread wheat confers heritable resistance to powdery mildew. *Nature Biotechnology*, *32*(9), 947–951.

Wenderoth, M., Pinecker, C., Voß, B., & Fischer, R. (2017). Establishment of CRISPR/Cas9 in Alternaria alternata. *Fungal Genetics and Biology*, *101*, 55–60.

Wilson, C. L., & Wisniewski, M. E. (1989). Biological control of postharvest diseases of fruits and vegetables: An emerging technology. *Annual Review of Phytopathology*, *27*(1), 425–441.

Wu, Z., Zhang, Y., Yu, H., Pan, D., Wang, Y., Wang, Y., Li, F., Liu, C., Nan, H., & Chen, W. (2021). Programmed genome editing by a miniature CRISPR-Cas12f nuclease. *Nature Chemical Biology*, *17*(11), 1132–1138.

Yang, X., Hameed, U., Zhang, A.-F., Zang, H.-Y., Gu, C.-Y., Chen, Y., & Xu, Y.-L. (2017). Development of a nested-PCR assay for the rapid detection of Pilidiella granati in pomegranate fruit. *Scientific Reports*, *7*(1), 1–8.

Yulianto, A. (2018). Pengaruh Literasi Keuangan Syariah Terhadap Keputusan Penggunaan Produk atau Layanan Lembaga Keuangan Syariah.

Zaynab, M., Sharif, Y., Fatima, M., Afzal, M. Z., Aslam, M. M., Raza, M. F., Anwar, M., Raza, M. A., Sajjad, N., & Yang, X. (2020). CRISPR/Cas9 to generate plant immunity against pathogen. *Microbial Pathogenesis*, *141*, 103996.

Zhou, X., Liao, H., Chern, M., Yin, J., Chen, Y., Wang, J., Zhu, X., Chen, Z., Yuan, C., & Zhao, W. (2018). Loss of function of a rice TPR-domain RNA-binding protein confers broad-spectrum disease resistance. *Proceedings of the National Academy of Sciences*, *115*(12), 3174–3179.

8 Bioethics and Risk Assessment of CRISPR-Edited Crops

INTRODUCTION

'An army marches on its stomach': Napoleon Bonaparte is reputed to have declared during the Battle of Waterloo, recognizing that several conflicts are won by those who have access to food supplies. The 'Great Famine' of Ireland, which lasted from the mid-1840s to the mid-1850s, resulted in the deaths of about 1 million people and the migration of over 2 million people (Grada, 1992). This is a striking illustration of how crop shortages may disrupt a whole country's demography.

Governments, scientists, and farmers are all working to boost agricultural production, as expected. Traditionally, breeding strategies have been beneficial in improving crop quantity and quality. In recent decades, crop yield has greatly increased with the use of fertilizers and pesticides. These compounds, on the other hand, lead to environmental pollution and may have a detrimental impact on food purity. An alternate solution and more eco-friendly techniques, such as genetic engineering, provide an easy and competent way for enhancing plant yields.

Modern genetic engineering is defined by the capacity to combine sequence specificity with the ability to generate DNA breaks. Modern methods include guided nucleases, which may cause DNA breaks and can be directed to a particular target location when combined with a desired DNA-binding domain (DBD). As a result, the host's DNA repair systems are stimulated, resulting in mutation and, perhaps, total loss of function of a specific gene of interest.

Meganucleases, zinc finger nucleases (ZFNs), transcription activator-like effector nucleases (TALENs), and CRISPR-Cas9 and orthologs are the four different kinds of designed nucleases that have been produced and exploited for genome engineering so far. All approaches have been used to edit genomes, but the first three have a restricted structure that is difficult to change, making them costly. Furthermore, instead of protein–DNA complexes, the system of CRISPR-Cas depends on guided-RNA (gRNA) and complexes of DNA–ribonucleotide for the identification of targets and consequent breaks. This ability also makes CRISPR more adaptable, simple to make, and less expensive (Gaj *et al.* 2013).

Genome modification may occur spontaneously, such as via mating or through genetic engineering utilizing genome-editing technologies (e.g., CRISPR-Cas9). When the DNA of an organism is altered by GE rather than naturally, it is referred to as a genetically modified organism (GMO). The CRISPR-Cas system is the most powerful genetic engineering technique presently accessible. This defensive mechanism

DOI: 10.1201/b22901-8

evolved in prokaryotes to protect them from viruses and other parasitic elements. Most of the archaea and several bacteria have CRISPR-Cas systems (Makarova *et al.* 2015; Burstein *et al.* 2016). The systems differ and, according to the most current categorization, there are two different classes, each of which is further split into six types (Koonin *et al.* 2017; Shmakov *et al.* 2017). CRISPR-Cas9 is the type II system that is mostly employed for genome editing nowadays. The mechanism was discovered to work as a guided-RNA enzyme that breaks DNA and causes double-strand breaks (DSBs) (Barrangou *et al.* 2007; Garneau *et al.* 2010; Gasiunas *et al.* 2012; Jinek *et al.* 2012). The structures of RNA, in which CRISPR RNA (CrRNA), trans-activating CRISPR RNA, and Cas9 are included, are required for the creation of effector complex and also used in the recognition of DNA in the system (Jinek *et al.* 2012).

A specific break in the target DNA is required for genome editing. Following the identification and binding of a particular target, CRISPR-Cas9 causes DSBs. Cas9 have two domains, RuvC and HNH, cause DSBs by cleaving the DNA at a precise place indicated by the guide RNA and leaving a blunt end (Jinek *et al.* 2012). Homology-directed repair (HDR) or nonhomologous end-joining (NHEJ) are the most frequent methods for repairing DSBs (Shrivastav *et al.* 2008). To repair DNA damage, HDR uses sequence homology. While NHEJ is an error-prone repair pathway.

Templates containing homologies around the cleaved area may be supplied to the cell with a double-strand DNA or single-strand DNA, causing accurate locus changes. On the other hand, NHEJ is the mechanism that is template independent, which results in mutation of gene insertion and/or deletion (indels). This mechanism changes the reading frame of genes or other critical regions such as promoters and other regulatory sequences (Lieber, 2010). This tool has been used to modify different organisms in several research studies.

Many industrial and academic research organizations have quickly accepted biotechnological advancements, such as using CRISPR-Cas for genetic modification. Indeed, the CRISPR-Cas editing technology has been rapidly adopted by the agricultural and food sectors (Belhaj *et al.* 2013; Liu *et al.* 2013). CRISPR is specifically employed to create genetically engineered plants, with the goal of significantly improving agricultural yields and other features seen in commercially available genetically modified crops. In addition, these are designed against various biotic and abiotic stresses (Castiglioni *et al.* 2008; Zaidi *et al.* 2016; Khatodia *et al.* 2017). Moreover, there are still several challenges in developing GE plants, especially securing regulatory and public opinions and permissions, and identifying whether they are GMOs or not. The concentration of this chapter will be on using CRISPR to create GM or non-GM plants, as well as the regulatory structure around GMO crops and the food they produce.

CRISPR-Cas FOR THE CREATION OF GMO PLANTS

Studies have shown that CRISPR-Cas9 technology can effectively modify the genomes of a wide range of crop species, like wheat, rice, tobacco, and tomato (Li *et al.* 2013; Nekrasov *et al.* 2013; Shan *et al.* 2013). Nonetheless, this showed the

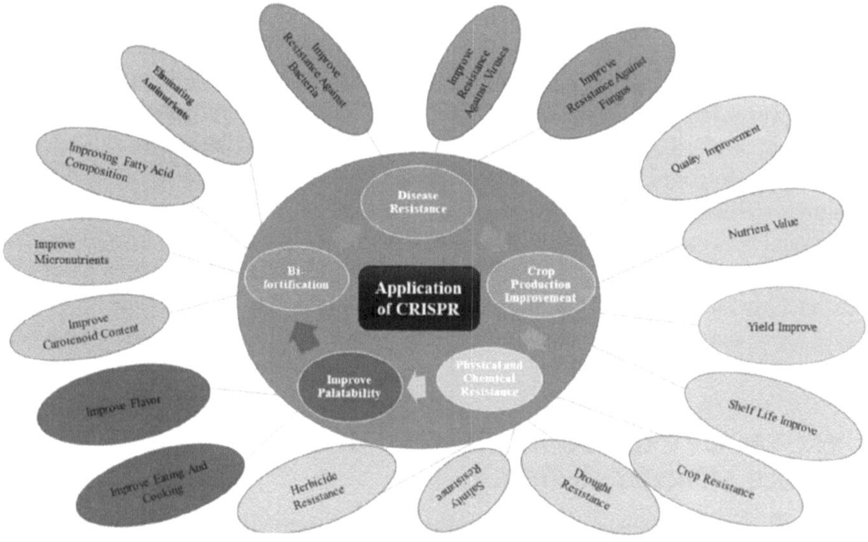

FIGURE 8.1 Applications of CRISPR-Cas system.

adaptability and promise of CRISPR-Cas technology for gene editing in plants, and that was further expanded and enhanced. For example, different vector systems have been developed to enhance CRISPR-Cas component expression and greater editing efficiency. In most of cases, many Cas9 genes have undergone codon optimization, which has resulted in increased efficiency (Xing *et al.* 2014). Additionally, the promoters that control vector expression which encodes the system of CRISPR have been changed. CRISPR system has been used for various purposes in plants, shown in (Figure 8.1).

Moreover, in the direction of developing GMOs for food production, introducing foreign DNA into plant cells raises regulatory hurdles. Additionally, Agrobacterium-mediated transformation is not the greatest effective strategy for all species of plants, and that's why many researchers revealed alternative delivery mechanisms (Woo *et al.* 2015).

The techniques of CRISPR-Cas have been used in the plant for many years and have been advancing all the time in targeting new targets and species. For example, CRISPR-Cas9 was used to target powdery mildew locus O (*MLO*) genes in tomatoes (Nekrasov *et al.* 2017). *MLO* genes code for proteins that make tomatoes susceptible to *Oidium neolycopersici*, a common fungal infection that causes powdery mildew disease (Acevedo-Garcia *et al.* 2014). *SlMl*ol was selected as a knockout target out of 16 *MLO* genes, since it is the greatest sponsor of the pathogen's susceptibility.

The altered plant was self-fertilized after PCR confirmed that the deletion had happened, resulting in offspring which did not have the DNA that encodes the cassette of CRISPR-Cas. As a consequence, resistance to the *Oidium neolycopersici* pathogen was proven in the transgene-free mutant plant with removals of one vulnerability gene. Because self-fertilization is not feasible in asexually reproducing plants, these techniques for creating transgene-free plants should not depend on a vector that is based on DNA to deliver the components of CRISPR to these plants. Kim et al. did not use a DNA cassette, rather pre-assembled CRISPR-Cas9 ribonucleoproteins (RNPs) were delivered into several plant species for targeted gene modifications (Woo *et al.* 2015). They isolated Cas9 proteins and combined them with sgRNA, targeting multiple genes from Arabidopsis, rice, and tobacco, similar to what they did in human cells and *Caenorhabditis elegans* (Cho *et al.* 2013; Kim *et al.* 2014). The target regions were examined for mutations after the complexes, constructed *in vitro*, were cultured with protoplasts obtained from these plants. In the examined plant species, success rates ranged from 8.4% to 44%. This method might alleviate regulatory issues about plant editing, meanwhile, no DNA is utilized to produce the construct.

APPROVAL OF CRISPR-ENGINEERED GMOs BY REGULATORY AGENCIES

Transgene-free methods for creating genetically modified organisms are beneficial as they avoid regulatory hurdles. Likewise, the United States Department of Agriculture (USDA) concluded in 2016 that genetically modified mushrooms created using the technique of CRISPR-Cas9 are excluded from the USDA's GMO laws (USDA–APHIS, 2016). In genes, the little deletions producing polyphenol oxidase—an enzyme essential in the mushroom's browning procedure were modified into white button mushrooms (*Agaricus bisporus*). The modifications decreased the enzyme's efficiency, giving the product a more attractive appearance and a longer lifespan. PCR and Western blot analyses indicated that no antibiotic markers were utilized for selection throughout the mushroom's development, and no remnants of the CRISPR-Cas9 system used during the procedure were remained. As a result, these mushrooms 'eluded' USDA restrictions, clearing the path for the introduction of more GMOs (Waltz, 2016).

In reaction to an investigation from the American corporation DuPont Pioneer, the USDA–APHIS wrote a letter along with the same terms (USDA–APHIS, 2016; Waltz, 2016). The *Wx1* gene, which encodes a synthase that produces the polysaccharide amylose, was knocked out in maize using CRISPR-Cas9. This change resulted in CRISPR-edited maize with starch forming a new polysaccharide called amylopectin, which is employed in several industries. According to the USDA's letter, this GMO also violates the agency's rules. Collectively, these instances show that regulatory agencies are willing to permit various GMO goods that were produced utilizing the CRISPR-Cas system, as far as the technology is deleted from the modified organism. This strategy makes sense since the finished result is equal to one created in the conventional manner, which would have required substantially more time, effort, and resources.

GLOBAL REGULATORY AUTHORITIES AND POLICIES FOR GMOs

The USDA is not the only entity in the United States that oversees GMOs. The Food and Drug Administration (FDA) and the Environmental Protection Agency (EPA) are also parts of the US regulatory structure (EPA). Depending on the GMO, one to three agencies may be necessary to approve it. The EPA, for example, regulates pesticide protection. Therefore, genetically modified species that are developed to resist specific insecticides are controlled by the EPA. Both the USDA and the FDA are expected to regulate products meant for human consumption (OSTP, 1986). The FDA and the USDA may have authority over the insertion of a gene expressing traits that assist drought tolerance, but the EPA does not.

The majority of US rules and policies are practical in nature and are dependent on the editing characteristic and the product's planned purpose (National Academies of Sciences, 2016). The primary considerations are safety and effectiveness, with cultural, moral, and social factors having less to no bearing. Furthermore, the FDA determined that there are not any significant dissimilarities among genetically altered and conventionally grown crops. As a result, the agency lacks the authority to impose any further labeling requirements (FDA, 2001).

The European Union (EU) entrusts scientific examination of GMO safety and environmental effects to the European Food Safety Authority (EFSA). In general, various member nations' scientific bodies collaborate with the EFSA to offer a scientific consensus. The European Commission makes the final decision. Some member states have strong public opposition to GM crops, making approval challenges. As a result, even if the European Commission approves a crop, it is unlikely that it will be grown in all of the EU member states. In the EU, genetically edited foods are required to be correctly labeled by law.

The variation in GMO approvals is reflected in the various rules and cultural views that govern different nations. Regulation and control of genetically edited crops and foods made from these crops follow distinct protocols. Governments' various methods reflect their differing reactions to public opinion and the scientific community. Different cultures, environmental circumstances, political interference, and interests of different groups like agricultural firms, and environmental activists or agencies are all reflected in the regulations.

The development of technology that allows gene editing without modifying the sequence of DNA, as well as GE plants without transgenes, has added to the complexity of regulatory rules. In principle, utilizing CRISPRi rather than traditional CRISPR-Cas9 to knock out a gene should make it simpler to circumvent regulatory constraints. Plants created in this manner are technically not GMOs, since the genome has not been altered. Moreover, in this context, the disadvantage of such approaches is that they just always need the existence and continual expression of CRISPR components. As a result of the technique, and particularly since they contain exogenous DNA, they are most likely to be classified as GMOs. Epigenome approaches, for example, might theoretically be produced transiently and then chosen against in offspring while keeping the alteration given. There is no precedent for this, as far as we know.

In general, these strategies still need to be improved in plants. The question regarding whether these plants are GMOs depends on if we look at the procedure or the final result (Araki and Ishii, 2015). Furthermore, several mutations induced by genome-editing tools that initiate the NHEJ pathway are far away from the USDA's regulatory reach, according to USDA judgments (Wolt *et al.* 2016). Collectively, these findings show that CRISPR technology may be utilized to create non-GMO plants.

Regulatory Policies for Genome Editing in the United States

The regulatory condition in the United States (US) is somewhat dissimilar, with biotechnology product regulation rules managed by a Coordinated Framework. The Environmental Protection Agency (EPA), The United States Department of Agriculture (USDA), and the Food and Drug Administration (FDA) are the key agencies involved in the organized framework for the evaluation of gene edited crops. These organizations deal with the environmental, agricultural, and human health consequences of bioengineered crops, correspondingly (Wolt, 2017).

The uses of CRISPR-Cas techniques do not need any new regulation, according to a 2016 USDA judgment, since the end item does not include foreign DNA from plant pests (Smyth, 2017; Waltz, 2016). As a result of this judgment, the culture of plants that was created using CRISPR-Cas techniques are exempt from the FDA's GMO restrictions, since they do not contain external DNA and cannot be discriminated from cultures created using conventional selections techniques (Ishii and Araki, 2017).

APHIS recommended SECURE Rule in May 2020 to regulate organisms developed using genetic engineering for plant pest risk with greater accuracy, precision and reduces regulatory burden for developers of organisms that are unlikely to pose plant pest risks. As a result, the United States principal regulatory organizations have yet to agree on whether the rules governing the health of people and ecological concerns should be applied to genetically edited organisms (Medvedieva and Blume, 2018). The USDA has announced that plants modified using the CRISPR-Cas system would not be regulated and that these plants may be produced and sold.

At least five CRISPR-Cas9 edited species have eluded the USDA's regulatory oversight in the past three years, including drought-tolerant soybean, browning-resistant mushrooms, high-amylopectin waxy corn (*Zea mays*) and false flax developed by USDA researchers (Waltz, 2016). The rules and policies in the United States are mostly technical in nature, and they depend on together designed qualities and the intended use of the goods (NASEM, 2016). Essentially, they examine the hazards presented by novel product characteristics within a synchronized framework. Other considerations like economical, ethical, and social concerns have little or no bearing on these rules and practices (Kleter *et al.* 2019).

The USDA does not regulate new kinds of agricultural plants created using CRISPRCas9 if they lack these features (Kleter *et al.* 2019). Whether plants developed using these approaches are classified as genetically modified depends on whether a procedure or a final item is studied. Prospective authorities and many researchers throughout the globe have urged that the regulation move away from a process-based approach and toward a product-based strategy (Globus and Qimron, 2018). Instead of assessing the procedure, which is time-consuming and costly, hazards could be assessed by looking at the final results after scanning off-target mutations. Moreover, the USDA views state that they do not control specific NHEJ pathway alterations generated by TALENs, ZFNs, meganucleases or CRISPR (Wolt *et al.* 2016). Collectively, these findings suggest that the techniques of CRISPR-Cas may be used to create plants that aren't classified as genetically edited in the United States.

Canadian Regulatory Process

Canada has created a biotechnology controlling framework centered on 'the product and its new characteristic', rather than the process employed to create the plant (Schuttelaar, 2015). Health Canada regulates the sale of foods made from plant novel characteristics (PNTs) via a premarket notice requirement (Wolt *et al.* 2016). As a result, CFIA and Health Canada authorized the herbicide-tolerant type of canola as the first commercial crop developed by gene editing in 2014 (Jones, 2015).

Genome-Editing Regulations in Europe

The complexities of the EU regulation for genetically modified plants have affected the protracted timeframe of regulatory and legal consideration for genetically modified plants. According to one understanding, these items should be controlled depending on the manufacturing procedure. Another viewpoint is that these items should be controlled in accordance with naturally produced products (Sprink *et al.* 2016). As a result, since they are dependent on the nature of the end product, it is necessary to develop regulatory regulations that are more practicable and scientifically reliable (Wolt *et al.* 2016). Regulators in Europe have mostly kept silent on gene-edited crops' final regulatory fates (Wolt, 2017).

The European Commission makes the ultimate judgment on whether or not to regulate genome-editing crops (EC). Even though crop certified by the European Commission is unlikely to be grown in all of the EU's member states, the significant public debate about genome-editing crops in certain of them precludes the EU from utilizing genetically edited crops (Globus and Qimron, 2018). Several EU member states, such as Sweden and Finland, have formed their own rules in favor of a non-genetically modified designation for genome-editing crops (Wolt, 2017) and did not wait for an official EU judgment (Spicer and Molnar, 2018).

Most of the nations that are pushing forward through issuing national suggestions have shown that they would be different from Europe Union decisions if they are issued. As a result, the idea of GMOs, as well as the hazards and restrictions connected with them, must be entirely redefined. For example, an upcoming characterization and risk valuation procedure might be collaborated on a certain kind of item or on a case-by-case basis, taking into consideration the benefits, drawbacks and dangers (Stilgoe *et al.* 2013). The 'innovation principle' may be used to gene-edited plants that are comparable to conventional types, predicting their potential ramifications for food security, environment protection, and economic troubles. Related hazards will be addressed on a case-by-case basis, allowing users to benefit from genetically edited items, though adhering to risk management guidelines.

As a result, it will create an appropriate strategy for regulating food security, environmental care, and the economy (Jouanin *et al.* 2018). Regulatory complications provide obstacles for genome-editing approaches, particularly those that use the removal of gene or nucleotide variations.

Furthermore, EASAC noted that rules' concerns must be directed at the applications rather than the gene-editing procedure as a new method (EASAC, 2017). The applications according to regulation must be based on evidence, take into account potential hazards and benefits, and be appropriate and flexible enough to accommodate future scientific breakthroughs. Furthermore, the EASAC urged EU authorities to confirm that genome-editing items that do not include DNA from a dissimilar organism are exempt from GMO regulations. Additionally, the EASAC advocates for total openness that shows the procedure utilized, and that the EU's goal should be to control a specific attribute or item rather than the technology used to generate it. These claims are in line with the exemption of genes that are newly edited techniques from restrictions if the genetic alterations caused by genome-editing methods are equivalent to traditional breeding items, and if no unique product-based danger is detected (Schulman *et al.* 2019). Plants created by genome editing could, in reality, be regulated similarly to plants produced through mutation breeding under a product-oriented regulatory framework (Jouanin *et al.* 2018).

Moreover, the ECJ has concluded that organisms/plants created using novel mutagenesis procedures should be controlled as GM goods in accordance with the language of Directive 2001/18/EC, 2001 (ECJ, 2018b; ECJ, 2018c). In reality, the use of approved genetically modified legislation in Europe to regulate genome-editing plants while at the same time exempting plants created via mutations in breeding is contradictory. Additional EU rules focused on the procedure utilized rather than the product created. The following cautious concept underpins this decision: 'A risk management method in which if a particular activity or policy has the potential to damage the public or ecologically, and there is not any scientific agreement on the problem, the rules or action in question cannot be implemented. In addition, scientific data becomes accessible, and the condition must be examined' (Jouanin *et al.* 2018). The ECJ considers organisms created by mutagenesis to be genetically edited, and, as a result, these creatures should comply with GM Directive's requirements.

The European Court of Justice believes that the hazards posed by new mutagenesis methods are comparable to those posed by transgenesis in GM plants. That's why exempting novel mutagenesis procedures from GM legislation's requirements would go against the GM Directive's goal and the cautious rules outlined above. The approach ECJ, which is based on process, considers the procedures used and the final items produced (Zannoni, 2019). Genetically modified rules in Europe for the editing of plant genomes are associated with costly ($35 million) and laborious (6 years) genetically edited care assessments and administrative procedures (Jouanin *et al.* 2018; McDougall, 2011) with unknown effects, since ultimate clearance is still a political choice. Furthermore, GM regulation removes the core advantages of genome editing—a low-cost, quick, and accurate technique for producing high-value-added plants that meet consumer and societal needs. Therefore, European researchers and firms are migrating to the US to conduct trials of genome editing (Burger and Evans, 2018).

As a result, labeling genetically modified plants would restrict European competitiveness, success, and revolutions to healthy foods (Jouanin *et al.* 2018). In February 2015, the House of Commons in the United Kingdom offered recommendations on the EU's GMO legislation's flaws, the implications of plant breeding methods in coming years, items derived from these techniques, and trade difficulties. The House of Commons stated that process-based restrictions had fallen behind the times and are possibly inhibiting innovation (House of Commons, 2015).

OBSTACLES AND SOLUTIONS FOR THE DEVELOPMENT OF CRISPR-BASED GMO

CRISPR-Cas Specificity

Precision is a key component of effective genome editing, especially when used to create GMOs, which need rigorous control. However, the mismatches that have occurred between single-guide RNA and target are controlled by CRISPR-Cas9, resulting in off-target editing (Fu *et al.* 2013). The delivery vectors may also be blamed for some of the side effects of CRISPR-Cas usage. For instance, these vectors were found not only to integrate into the host's genome but also mistakenly used by host cells as templates for the repair of DNA break (Kim *et al.* 2014).

GMOs with off-target mutations can be detected using whole-genome sequencing or Di-genome sequencing (Kim *et al.* 2015; Kim and Kim, 2016). These techniques confirm that just the necessary alterations have been made to the GMO. To limit off-targets, a third strategy is to choose much-specified target sequences. This is accomplished via the use of basic and easily available techniques like BLAST or particular databases (Xie *et al.* 2014). Besides, the shortening of gRNA can also dramatically minimize off-target activity (Dang *et al.* 2015; Ran *et al.* 2013; Slaymaker *et al.* 2016).

Public Lack of Acceptance of GMOs

The lack of public support for genetically edited organisms is a key barrier to their adoption and growth. An unfavorable attitude regarding GMOs is often linked to a lack of understanding and a lack of faith in developers and regulators (Siegrist *et al.* 2012; Lucht, 2015). It is thus critical to enhance public knowledge by informing people about the various breeding procedures and technology used to create GMOs.

Consumers may make educated judgments based on their reasonable judgment when there is more openness. Transparency will also help to promote science and gain public acceptance. As a result, there would be more product demand, which would motivate biotechnology scientists to engage in the development of research. Though cultural and social aspects are important, rational decision-making requires a thorough grasp of technology and the science underlying it. In this perspective, the important results of a study by the National Academies of Sciences, Engineering, and Medicine on the detrimental and good consequences of genetically modified crops should be mentioned (National Academies of Sciences, 2016). According to the paper, there is essentially no evidence that genetically modified crops have harmful environmental consequences.

On the other hand, several advantages were discovered, such as decreased pesticide exposure. In the case of people, there was little verification to suggest any dangers. Animal studies have not found any distinguishing characteristics in health between genetically modified and nongenetically modified diets. However, the committee acknowledged that certain risk assessment methodologies are restricted and that long-term consequences are difficult to monitor and understand, both in terms of the environment and human health.

Ethical Concerns

Despite the accessibility of CRISPR-Cas technology for genetically editing, the cost of producing GMOs remains high. Producing, testing, and bringing novel crops to the market generally requires a significant amount of cash investment as well as the support of huge corporations or organizations. As a result, society must consider how GMO crops may hurt small-scale farmers who are unable to compete with more costly proprietary strains with greater yields. Increased yields, drought and disease resistance, and plants modified to generate extra vitamins, among other things, might help developing nations with a dearth of crops or a shortage of certain critical elements in the crops they cultivate. They might also profit from crops that are less contaminated by fungal toxins, which is a common problem in poor nations (Wild and Gong, 2010).

Agriculture is important in developing nations, although it is mostly focused on Western countries economically. Because the primary debates on GMO research take place mostly in the United States and the European Union, the enormous advantages that GMOs may bring to poor nations are often overlooked. When making GMO judgments, this concern should also be taken into account. The difficulties raised here aren't entirely related to CRISPR, and the majority of them predate the technology. Since the introduction of GE plants, ethical concerns, public opposition, how to govern GMOs, and even what constitutes a GMO have all been raised.

SUMMARY AND PROSPECTS

Because of its durability, effectiveness, and broad variety of uses, CRISPR has become an important gene-editing technique in recent years. While, opening the path for more accurate cellular research, CRISPR technology is making it simpler to modify single genes or nucleotides quickly and correctly, CRISPR could serve to efficiently activate or silence genes of interest, aiding in the elucidation of their functions in particular pathways (Qi *et al.* 2013) or it can be used to resensitize drug-resistant bacteria to antibiotics or to prevent spread of bacterial resistance. CRISPR-based technology is increasingly phasing out the use of transgenic DNA in plants. It also achieves exact and specific verified alterations of the target gene without causing any side effects, which alleviates regulatory worries. CRISPR technology's accessibility is facilitating a growing quantity of research. Despite the fact that CRISPR was just discovered a decade ago, there is an abundance of data and tools available. It's simple to imagine that in the near future, many more breakthroughs and unique uses will be observed. Regulatory choices should be founded on transparent scientific facts, therefore ethical considerations should be explored, and this technique must be exposed to the public as much as conceivable.

REFERENCES

Acevedo-Garcia, J., Kusch, S., & Panstruga, R. (2014). Magical mystery tour: MLO proteins in plant immunity and beyond. *New Phytologist, 204*:273–281.

Ali, Z., Abul-faraj, A., & Li, L. (2015). Efficient virus-mediated genome editing in plants using the CRISPR/Cas9 system. *Molecular Plant*, *8*:1288–1291.

Araki, M. & Ishii, T. (2015). Towards social acceptance of plant breeding by genome editing. 2015. *Trends in Plant Science*, *20*: 145–149.

Barrangou, R. & Doudna, J.A. (2016). Applications of CRISPR technologies in research and beyond. *Nature Biotechnology*, *34*:933–941.

Barrangou, R., Fremaux, C., & Deveau, H. (2007). CRISPR provides acquired resistance against viruses in prokaryotes. *Science*, *315*:1709–1712.

Belhaj, K., Chaparro-Garcia, A., Kamoun, S., & Nekrasov, V. (2013). Plant genome editing made easy: Targeted mutagenesis in model and crop plants using the CRISPR/Cas system. *Plant Methods*, *9*:39.

Brooks, C., Nekrasov, V., Lippman, Z.B., & Van Eck, J. (2014). Efficient gene editing in tomato in the first generation using the clustered regularly interspaced short palindromic repeats/CRISPR-associated system. *Plant Physiology*, *166*:1292–1297.

Burstein, D., Harrington, L.B., & Strutt, S.C. (2016). New CRISPR-Cas systems from uncultivated microbes. *Nature*, *542*:237–241.

Castiglioni, P., Warner, D., & Bensen, R.J. (2018). Bacterial RNA chaperones confer abiotic stress tolerance in plants and improved grain yield in maize under water-limited conditions. *Plant Physiology*, *147*:446–455.

Cho, S.W., Lee, J., Carroll, D., Kim, J.S., & Lee, J. (2013). Heritable gene knockout in Caenorhabditis elegans by direct injection of Cas9-sgRNA ribonucleoproteins. *Genetics*, *195*:1177–1180.

Cong, L., Ran, F.A., & Cox, D. (2013). Multiplex genome engineering using CRISPR/Cas systems. *Science*, *339*:819–823.

Dang, Y., Jia, G., & Choi, J. (2015). Optimizing sgRNA structure to improve CRISPR Cas9 knockout efficiency. *Genome Biology*, *16*:280.

DiCarlo, J.E., Norville, J.E., Mali, P., Rios, X., Aach, J., & Church, G.M. (2013). Genome engineering in Saccharomyces cerevisiae using CRISPR-Cas systems. *Nucleic Acids Research*, *41*:4336–4343.

Esvelt, K.M., Mali, P., Braff, J.L., Moosburner, M., Yaung, S.J., & Church, G,M. (2013). Orthogonal cas9 proteins for RNA-Guided gene regulation and editing. *Nature Methods*, *10*:1116–1121.

Fauser, F., Schiml, S., & Puchta, H. (2014). Both CRISPR/Cas-based nucleases and nickases can be used efficiently for genome engineering in Arabidopsis thaliana. *The Plant Journal*, *79*:348–359.

FDA (U.S. Department of Health and Human Services–Food and Drug Administration). (2001). Premarket Notice Concerning Bio- engineered Foods; Proposed Rule. *Federal Register, 66*:4706–4738.

Feng, Z., Zhang, B., Ding, W., et al. (2013). Efficient genome editing in plants using a CRISPR/Cas system. *Cell Research*, *23*:1229–1232.

Friedland, A.E., Tzur, Y.B., Esvelt, K.M., Colaiacovo, M.P., Church, G.M., & Calarco, J.A. (2013). Heritable genome editing in C. elegans via a CRISPR- Cas9 system. *Nature Methods*, *10*:741–743.

Fu, Y., Foden, J.A, Khayter, C., et al. (2013). High-frequency off-target mutagenesis induced by CRISPR-Cas nucleases in human cells. *Nature Biotechnology*, *31*:822–826.

Gaj, T., Gersbach, C.A., & Barbas, C.F. (2013). ZFN, TALEN, and CRISPR/ Cas-based methods for genome engineering. *Trends in Biotechnology*, *31*:397–405.

Gao, Y. & Zhao, Y. (2014). Self-processing of ribozyme-flanked RNAs into guide RNAs in vitro and in vivo for CRISPR-mediated genome editing. *Journal of Integrative Plant Biology*, *56*:343–349.

Garneau, J.E., Dupuis, M.È., Villion, M., et al. (2010). The CRISPR/Cas bacterial immune system cleaves bacteriophage and plasmid DNA. *Nature*, *468*:67–71.

Gasiunas, G., Barrangou, R., Horvath, P., & Siksnys, V. (2012). Cas9-crRNA ribonucleoprotein complex mediates specific DNA cleavage for adaptive immunity in bacteria. *Proceedings of the National Academy of Sciences of the United States of America*, *109*: E2579–E2586.

Gilbert, L.A., Larson, M.H., Morsut, L., et al. (2013). CRISPR-mediated modular RNA-guided regulation of transcription in eukaryotes. *Cell*, *154*:442–451.

Gratz, S.J., Cummings, A.M., Nguyen, J.N., et al. (2013). Genome engineering of drosophila with the CRISPR RNA-Guided cas9 nuclease. *Genetics*, *194*:1029–1035.

Hilton, I.B., D'Ippolito, A.M., Vockley, C.M., et al. (2015). Epigenome editing by a CRISPR-Cas9-based acetyltransferase activates genes from promoters and enhancers. *Nature Biotechnology*, *33*:510–517.

Hsu, P.D., Scott, D.A., Weinstein, J.A., et al. (2013). DNA targeting specificity of RNA-guided Cas9 nucleases. *Nature Biotechnology*, *31*:827–832.

Hwang, W.Y., Fu, Y., Reyon, D., et al. (2013). Efficient genome editing in zebrafish using a CRISPR-Cas system. *Nature Biotechnology*, *31*:227–229.

Ibrahim, M.A., Griko, N., Junker, M., & Bulla, L.A. (2010). *Bacillus thuringiensis*. *Bioengineered Bugs*, *1*:31–50.

Jiang, W., Bikard, D., Cox, D., Zhang, F., & Marraffini, L.A. (2013a). RNA-guided editing of bacterial genomes using CRISPR-Cas systems. *Nature Biotechnology*, *31*:233–239.

Jiang, W., Zhou, H., Bi, H., Fromm, M., Yang, B., & Weeks, D.P. (2013b). Demonstration of CRISPR/Cas9/sgRNA-mediated targeted gene modification in arabidopsis, tobacco, sorghum, and rice. *Nucleic Acids Research*, *41*: e188.

Jinek, M. (2012). Programmable dual-RNA-guided DNA endonuclease in adaptive bacterial immunity. *Science*, *337*:816–821.

Kearns, N.A., Pham, H., Tabak, B., et al. (2015). Functional annotation of native enhancers with a Cas9-histone demethylase fusion. *Nature Methods*, *12*:401–403.

Khatodia, S., Bhatotia, K., & Tuteja, N. (2017). Development of CRISPR/Cas9 mediated virus resistance in agriculturally important crops. *Bioengineered*, *8*:274–279.

Kim, D., Bae, S., Park, J., et al. (2015). Digenome-seq: Genome-wide profiling of CRISPR-Cas9 off-target effects in human cells. *Nature Methods*, *12*:237–243.

Kim, J. & Kim, J.-S. (2016). Bypassing GMO regulations with CRISPR gene editing. *Nature Biotechnology, 34*:1014–1015.

Kim, S., Kim, D., Cho, S.W., Kim, J., & Kim, J.S. (2014). Highly efficient RNA- guided genome editing in human cells via delivery of purified Cas9 ribonucleoproteins. *Genome Research*, *24*:1012–1019.

Kiro, R., Shitrit, D., & Qimron, U. (2014). Efficient engineering of a bacteriophage genome using the type I-E CRISPR-Cas system. *RNA Biology, 11*:42–44.

Koonin, E.V., Makarova, K.S., & Zhang, F. (2017). Diversity, classification, and evolution of CRISPR-Cas systems. *Current Opinion in Microbiology*, *37*: 67–78.

Li, J.F., Norville, J.E., Aach, J., et al. (2013). Multiplex and homologous recombination-mediated genome editing in Arabidopsis and *Nicotiana benthamiana* using guide RNA and Cas9. *Nature Biotechnology, 31*:688–691.

Lieber, M.R. (2010). The mechanism of double-strand DNA break repair by the nonhomologous DNA end-joining pathway. *Annual Review of Biochemistry*, *79*:181–211.

Liu, W., Yuan, J.S., & Stewart, C.N., Jr. (2013). Advanced genetic tools for plant biotechnology. *Nature Reviews Genetics*, *14*:781–793.

Liu, Y., Zhan, Y., Chen, Z., et al. (2016). Directing cellular information flow via CRISPR signal conductors. *Nature Methods, 13*:938–944.

Lo, T.W., Pickle, C.S., Lin, S., et al. (2013). Precise and heritable genome editing in evolutionarily diverse nematodes using TALENs and CRISPR/ Cas9 to engineer insertions and deletions. *Genetics*, *195*: 331–348.

Lowder, L.G., Zhang, D., Baltes, N.J., et al. (2015). A CRISPR/Cas9 toolbox for multiplexed plant genome editing and transcriptional regulation. *Plant Physiology*, *169*:971–985.

Lucht, J. (2015). Public acceptance of plant biotechnology and GM crops. *Viruses*, *7*:4254–4281.

Makarova, K.S., Wolf, Y.I., Alkhnbashi, O.S., et al. (2015). An updated evolutionary classification of CRISPR–Cas systems. *Nature Reviews Microbiology*, *13*:722–736.

Mali, P., Aach, J., Stranges, P.B., et al. (2013a). CAS9 transcriptional activators for target specificity screening and paired nickases for cooperative genome engineering. *Nature Biotechnology, 31*:833–838.

Mali, P., Yang, L., Esvelt, K.M., et al. (2013b). RNA-guided human genome engineering via cas9. *Science, 339*:823–826.

Mao, Y., Zhang, H., Xu, N., Zhang, B., Gou, F., & Zhu, J.K. (2103). Application of the CRISPR-Cas system for efficient genome engineering in plants. *Molecular Plant, 6:* 2008–2011.

Martel, B. & Moineau, S. (2104). CRISPR-Cas: An efficient tool for genome engineering of virulent bacteriophages. *Nucleic Acids Research, 42:* 9504–9513.

Miao, J., Guo, D., Zhang, J., et al. (2013).Targeted mutagenesis in rice using CRISPR-Cas system. *Cell Research, 23*:1233–1236.

National Academies of Sciences, Engineering, Medicine. (2016). *Genetically Engineered Crops: Experiences and Prospects*. The National Academies Press. pp. 98–237.

Nekrasov, V., Staskawicz, B., Weigel, D., Jones, J.D.G., & Kamoun, S. (2013). Targeted mutagenesis in the model plant *Nicotiana benthamiana* using Cas9 RNA-guided endonuclease. *Nature Biotechnology, 31*:691–693.

Nekrasov, V., Wang, C., Win, J., Lanz, C., Weigel, D., & Kamoun, S. (2017). Rapid generation of a transgene-free powdery mildew resistant tomato by genome deletion. *Science Reports, 7*:482.

Ó Gráda C. (1992). Ireland's great famine. Recent Findings of Research in Economic & Social History. Dublin, Ireland: University College Dublin; Centre for Economic Research. 5–8.

OSTP (Executive Office of the President, Office of Science and Technology Policy). (1986). Coordinated Framework for Regulation of Biotechnology. Federal Register 51:23302. Available at https:// www.aphis.usda.gov/brs/fedregister/coordinated_framework.pdf.

Piatek, A., Ali, Z., Baazim, H., et al. (2015). RNA-guided transcriptional regulation in planta via synthetic dCas9-based transcription factors. *Plant Biotechnology Journal, 13*:578–589.

Polstein, L.R. & Gersbach, C.A. (2015). A light-inducible CRISPR-Cas9 system for control of endogenous gene activation. *Nature Chemistry Biology, 11*: 198–200.

Qi, L.S., Larson, M.H., Gilbert, L.A., et al. (2013). Repurposing CRISPR as an RNA-Guided platform for sequence-specific control of gene expression. *Cell, 152*:1173–1183.

Ran, F.A., Hsu, P.D., Lin, C.-Y., et al. (2013). Double nicking by RNA-guided CRISPR cas9 for enhanced genome editing specificity. *Cell, 154*:1380–1389.

Shan, Q., Wang, Y., Li, J., et al. (2013). Targeted genome modification of crop plants using a CRISPR-Cas system. *Nature Biotechnology, 31*: 686–688.

Sherkow, J.S. (2017). Inventive steps: The CRISPR patent dispute and scientific progress: The recent patent decisions about CRISPR tell us a lot about how advances in biology are actually made-and how they are not. *EMBO Reports, 18*:1047–1051.

Shmakov, S., Smargon, A., Scott, D., et al. (2017). Diversity and evolution of class 2 CRISPR-Cas systems. *Nature Reviews Microbiology, 15*: 169–182.

Shrivastav, M., De Haro, L.P., & Nickoloff, J.A. (2008). Regulation of DNA double-strand break repair pathway choice. *Cell Research, 18*: 134–147.

Siegrist, M., Connor, M., & Keller, C. (2012). Trust, confidence, procedural fairness, outcome fairness, moral conviction, and the acceptance of GM field experiments. *Risk Analysis, 32*:1394–1403.

Slaymaker, I.M., Gao, L., Zetsche, B., Scott, D.A., Yan, W.X., Zhang, F. (2016). Rationally engineered Cas9 nucleases with improved specificity. *Science, 351*:84–88.

USDA–APHIS (U.S. Department of Agriculture–Animal and Plant Health Inspection Service). (2016a). Confirmation of Regulatory Status of Waxy Corn Developed by CRISPR-Cas Technology. Available at https://www.aphis.usda.gov/biotechnology/downloads/reg_loi/15–352-01_air_response_signed.pdf.

USDA–APHIS (U.S. Department of Agriculture–Animal and Plant Health Inspection Service). (2016b). Request for confirmation that transgene-free CRISPR-edited mushroom is not a regulated article. https://www.aphis.usda.gov/biotechnology/downloads/reg_loi/15-32101_air_response_signed.pdf.

Waltz, E. (2016a). CRISPR-edited crops free to enter market, skip regulation. *Nature Biotechnology, 34*:582.

Waltz, E. (2016b). Gene-edited CRISPR mushroom escapes US regulation. *Nature, 532*:293–293.

Wang, H., La Russa, M., & Qi, L.S. (2016). CRISPR/Cas9 in genome editing and beyond. *Annual Review of Biochemistry, 85*:227–264.

Wang, H., Yang, H., Shivalila, C.S., et al. (2013). One-step generation of mice carrying mutations in multiple genes by CRISPR/Cas-mediated genome engineering. *Cell, 153*:910–918.

Wechsler, S.J. (2016). Recent trends in GE adoption. Available at https:// www.ers.usda.gov/data-products/adoption-of-genetically-engineered-crops-in-the-us/recent-trends-in-geadoption/. Accessed. 2:2017.

Wild, C.P. & Gong, Y.Y. (2010). Mycotoxins and human disease: A largely ignored global health issue. *Carcinogenesis, 31*:71–82.

Wolt, J.D., Wang, K., & Yang, B. (2016). The regulatory status of genome-edited crops. *Plant Biotechnology Journal, 14*:510–518.

Woo, J.W., Kim, J., Kwon, S.I., et al. (2015). DNA-free genome editing in plants with preassembled CRISPR-Cas9 ribonucleoproteins. *Nature Biotechnology, 33*:1162–1164.

Wright, A.V., Nuñez, J.K., & Doudna, J.A. (2016). Biology and applications of CRISPR systems: Harnessing nature's toolbox for genome engineering. *Cell, 164*:29–44.

Xie, K. & Yang, Y. (2013). RNA-guided genome editing in plants using a CRISPR-Cas system. *Molecular Plant, 6*:1975–1983.

Xie, K., Zhang, J., & Yang, Y. (2014). Genome-wide prediction of highly specific guide RNA spacers for CRISPR-Cas9-Mediated genome editing in model plants and major crops. *Molecular Plant, 7*:923–926.

Xing, H.-L., Dong, L., Wang, Z.-P., et al. (2014). A CRISPR/Cas9 toolkit for multiplex genome editing in plants. *BMC Plant Biology, 14*:327.

Yao, J., Weng, Y., Dickey, A., & Wang, K.Y. (2015). Plants as factories for human pharmaceuticals: Applications and challenges. *International Journal of Molecular Sciences, 16*:28549–28565.

Yosef, I., Manor, M., Kiro, R., & Qimron, U. (2015). Temperate and lytic bacteriophages programmed to sensitize and kill antibiotic-resistant bacteria. *Proceedings of the National Academy of Sciences of the United States of America, 112*:7267–7272.

Zaidi, S.S.A., Tashkandi, M., Mansoor, S., & Mahfouz. M.M. (2016). Engineering plant immunity: Using CRISPR/Cas9 to generate virus resistance. *Front Plant Science, 7*:1673.

Zhou, H., Liu, B., Weeks, D.P., Spalding, M.H., & Yang, B. (2014). Large chromosomal deletions and heritable small genetic changes induced by CRISPR/Cas9 in rice. *Nucleic Acids Research, 42*:10903–10914.

Index

Note: **Bold** page numbers refer to tables and *italic* page numbers refer to figures.